Studies in Ecological Economics

Volume 7

The series "Ecological Economics" publishes peer reviewed monographs and edited volumes that offer fresh perspectives and overviews on a range of topics in Ecological Economics. Special attention will be given to integrative research methods and techniques and to pluralistic approaches. Fundamental and critical discussions are invited of: theoretical assumptions, ethical starting points, behavioural models, (co)evolutionary change, integration concepts, the ecosystem approach and ecosystem services environmental macroeconomics, industrial ecology, spatial dimensions, thermodynamics and production functions, policy goals and instruments, international policy dimensions, alternative welfare measures, valuation and benefits transfer. Particular emphasis will be placed on the interaction between valuation, modelling and evaluation in a multidisciplinary setting; the link between ecology, biodiversity, ecosystem services, economics politics and environmental management; the incorporation of physical flows in economic models; and the interface between development, poverty, technology and sustainability. In addition, applied and policy oriented research is welcomed, addressing specific resources, substances, materials, regions, sectors, countries or environmental problems. International comparative studies are also encouraged. The ultimate aim of this series is to present a rigorous but broad perspective on contemporary and future environmental policy questions.

The series welcomes new proposals! To submit a proposal, please fill in a "book proposal form" available on the series page, or contact juliana.pitanguy@springer.com.

Dipayan Dey • Joyashree Roy
Editors

Biorights

The Neo-economic Conservation Paradigm for Global South

Springer

Editors
Dipayan Dey
South Asian Forum for Environment
Kolkata, West Bengal, India

Joyashree Roy
Asian Institute of Technology
Bangkok, Thailand

Asia Pacific Network for Global Change Research, Japan

ISSN 1389-6954 ISSN 2542-9531 (electronic)
Studies in Ecological Economics
ISBN 978-3-030-91502-5 ISBN 978-3-030-91503-2 (eBook)
https://doi.org/10.1007/978-3-030-91503-2

This Springer imprint is published by the registered company Springer Nature Switzerland AG
The registered company address is: Gewerbestrasse 11, 6330 Cham, Switzerland

Preface

Economics of environmental services has been emerging as an academic discipline for many decades now. However, as an operational guiding principle, this is becoming as an essential domain of knowledge both for academia as well as among the practitioners to resolve the problem of allocating the monetary values to non-marketed nature's goods and services including biodiversity. However, in the 'Climate Change' discourse, the significance has been felt more urgently as the debate around role of ecosystem/natural capital is gaining more and more prominence due to accelerated action towards global sustainability. Both adaptive and mitigative actions are evaluating the potential for ecosystem-based solutions.

Academia has systematically developed scientific arguments and logical framework to expand the concept of capital as a major driver of economic growth through portfolio approach to cover natural capital among other forms of capital to represent role of environmental services in context of economic growth. So, equity and inequity in capital ownership have also been extended to the domains of natural capital keeping in mind the inclusion in decision-making process and empowerment. This has been used to address several aspects and issues of socioecological and socioeconomic significance like gender inequity, equitable access to resources, traditional and indigenous knowledge, community preparedness, and women empowerment in designing sustainable development models. Community rights on natural capital and its implications on natural resource management/transfer of rights is one such area of research that tries to enunciate the pros and cons of community-led conservation and its rights to access nature's goods and services, as well from integrative aspect to macroeconomic perspective.

While there is urgency to accelerate the efforts towards the net zero goal by mid-century to ensure long-term sustainability, the first stepping stone on that trajectory is through attainment of sustainable development goals (SDGs) by 2030. The urgency therein and the opportunities that have opened up new dimensions in climate and non-climate financing and also orchestrate through minimisation the trade-offs and enhancement of synergies at community-ecosystem interface. The most important issues pertaining to justice, fairness and equity perspective are the rights and responsibilities of vulnerable communities living in the margins of dominant market-based

Gross Domestic Product (GDP)-centric economic growth-dominated socio-economic system that are engaged in conservation of environment and biodiversity intergenerationally as livelihood option but are also identified as vulnerable communities with high exposure to climate variability and change due to sole dependence for livelihood extensively dependent on nature's goods and services.

In this context, we juxtapose the issue of biorights. Some of the frequently asked questions are: how to address the right of marginal communities which has originated through their intergenerational presence and intertwined-ness through interaction and interdependence on natural capital/ecosystem historically conferred custodian intergenerational? How to integrate this non-market institution-based relation with nature and time invested by the communities on conservation of that ecosystem with mainstream, market-based property right socio-economic regimes? In this context, neo-economic conservation paradigm like biorights going beyond market as an institution comes in. Biorights is not an entirely new concept. It existed in many primordial forms and across traditional practices in the communities. These practices prompted use of resources which were considered wise to sustain the communities' livelihood intergenerationally through the ages. However, how these traditional practices can inform various expansion of solution spaces in modern market-driven socio-economic systems has been addressed in this book. It explores the historically acquired rights through practices and how they are negotiating within post industrial revolution legal structures to represent the genesis of biorights and understand its significance in the context of climate change. It prompts as well the question around economic instruments in establishing biorights of the communities keeping within the statutory frameworks of societal, economic and legal post industrial revolution regime. Most of the discussions and deliberations in the chapters have been annotated with examples, empirical evidence from the context-specific field-based case studies, interventions, and community practices in the context of Asian region. This would help the readers understand the concept of biorights in a more bottom-up, pragmatic way and build on the actual context, voices of the people, as well as within a logical and reasonable interpretation in larger contexts.

We desire the book finds its place in appropriate hands: of academia, researchers and practitioners, policy planners, and community leaders. Suggestions, reviews and feedbacks are highly welcome so that they can help us improve it further for better value and use at all levels.

Kolkata, India
September 2021

Dipayan Dey
Joyashree Roy

Acknowledgement

We are thankful to the Asia-Pacific Network for Global Change Research for supporting our studies and research, which has helped us in designing this publication. We express our thanks to all the contributing authors for their valued submissions. We are also thankful to the South Asian Forum for Environment [SAFE] for sharing knowledge and evidences from their field interventions on biorights, which has guided us to develop the concept to a presentable framework.

Last but not least, we are thankful to all our friends and families, without whose support and motivation this manuscript would not have been a successful publication.

Kolkata, India
September 2021

Dipayan Dey
Joyashree Roy

Contents

Contributors

Amitava Aich South Asian Forum for Environment, Kolkata, West Bengal, India

Priyanie Amerasinghe International Water Management Institute, Colombo, Sri Lanka

Udo Censkowsky Blue Census, Weilheim, Germany

Dipayan Dey South Asian Forum for Environment, Kolkata, West Bengal, India

Malancha Dey Roy South Asian Forum for Environment, Kolkata, West Bengal, India

Sourabh Kumar Dubey Blue Census, Weilheim, Germany

Ramchandran Gopichandran NTPC School of Business, Noida, India

NCR, Delhi, India

Joyashree Roy Asian Institute of Technology, Bangkok, Thailand

Jadavpur University, Kolkata, India

Abbreviations

BPL	Below Poverty Line
CAGR	Compound Annual Growth Rate
CBD	Convention on Biological Diversity
CBFM	Community Based Forest Management
CBNRM	Community Based Natural Resource Management
CDMs	Clean Development Mechanisms
CEDAW	Convention on the Elimination of All Forms of Discrimination against Women
COP	Convention on Climate Change
CSO	Civil Society Organization
CTI	Clean Technology Initiatives
DBW	Deepor Beel Wetland
DFID	Department for International Development
EbA	Ecosystem-Based Adaptation
EEZ	Economic Environmental Zones
EKW	East Kolkata Wetland
ENSO	El Nino Southern Oscillation
ESs	Ecosystem Services
ESCAP	Economic and Social Commission for Asia and the Pacific
ESRC	Economic and Social Research Council
FSC	Forest Stewardship Council
FPIC	Free, Prior, and Informed Consent
GDP	Gross Domestic Product
ICCAs	Indigenous and Community Conserved Areas
ICCPR	International Covenant on Civil and Political Rights
ICDP	Integrated Conservation and Development Projects
ICESCR	International Covenant on Economic, Social and Cultural Rights
ICS	Internal Control System
ILO	International Labour Organization
IMA	Integrated Mangrove Aquaculture

IMCs	Indian Major Carps
IMFFS	Integrated Mangrove Fishery Farming System
IOD	Indian Ocean Dipole
IPCC	Intergovernmental Panel on Climate Change
IUCN	International Union for Conservation of Nature
JFMCs	Joint Forest Management Committees
JLGs	Joint Liability Groups
LBPL	Lower Bound Poverty Line
LNOB	Leave No One Behind
LULC	Land Use and Land Cover
MEA	Millennium Ecosystem Assessment
MSMEs	Micro, Small and Medium Enterprises
MSSRF	M. S. Swaminathan Research Foundation
NAMAs	National Appropriate Mitigation Actions
NBCC	North Borneo Chartered Company
NbS	Nature-based Solutions
NERC	Natural Environment Research Council
NGO	Non-governmental Organization
NPOP	National Programme for Organic Production
NWP	National Wetland Policy
OAPS	Organic Agricultural Products
OHCHR	Office of the High Commissioner for Human Rights
PES	Payments from Environmental Services
PFES	Payment for Forest Environmental Services
PPF	Public Provident Fund
PRESA	Pro-Poor Rewards for Environmental Services in Africa Programme
REDD	Reducing Emissions from Deforestation and Forest Degradation
RTPCR	Reverse Transcription Polymerase Chain Reaction
SDGs	Sustainable Development Goals
SEPLs	Socioecological Production Landscape
SFD	Sabah Forestry Department
SFM	Sustainable Forest Management
SFMLA	SFM License Agreement
SOC	Soil Organic Carbon
SOP	Standard Operating Protocol
TEEB	The Economics of Ecosystems and Biodiversity
TFDP	Third Forestry Development Project
TIES	The International Ecotourism Society
UNDRIP	UN Declaration on the Rights of Indigenous People
UNEP	United Nations Environment Programme
UNFCCC	United Nations Framework Convention on Climate Change
UTs	Union Territories
WEF	World Economic Forum
WSSV	White Spot Syndrome Virus

Chapter 1
Introduction to Biorights

Malancha Dey Roy

1.1 Nature's Services in Primitive Agrarian Societies

Science of sustainability is of utmost concern in today's world facing the burden of climate crisis, though its handprints are well available since primitive civilization like Indus Valley or Mesopotamian societies were in existence (Iqbal, 2017; Kineman & Anand, 2014). They created a sustainable and inter-reliant balance amongst economic, social and ecological development. Agriculture was the prime means of earning for livelihood and the agrarian communities conglomerated around suitable agro-ecologies, especially riverine fertile floodplains. In the course of management of natural resources for agro-farming practices like usage of irrigation water, draught animals and other associated paraphernalia, proclivity of responsiveness towards nature's goods and services and as well, the community rights on these basic amenities for farming, gradually spawned as a societal conditioning in the agrarian communities. The 'Neolithic Revolution' had triggered a change in societal practices and lifestyles to ensure a consistent food supply. Factors relating to differential climatic change in various parts of the world determined the pattern and methodology of agricultural practices in the locale (Zhang et al., 2017). Gradual changes in ancient cultivation techniques, controlled use of natural resources, transition from wild harvesting to traditional crop and cereal culture, domestication of plants, livestock rearing was indicator of the shift from an itinerant to a stable way of life. Civilizations may have been made feasible by agriculture, but it has never been a shield against their fall over space and time. The fall, in fact, led to the origin of more resilient forms of settlements. Increases in agricultural production had struggled against population expansion, resource depletion, droughts, changing climates, and other challenges throughout the civilization. Degradation of natural

M. D. Roy (✉)
South Asian Forum for Environment, Kolkata, West Bengal, India
e-mail: research@safeinch.org

D. Dey, J. Roy (eds.), *Biorights*, Studies in Ecological Economics 7,
https://doi.org/10.1007/978-3-030-91503-2_1

resources, depleted farmlands, changing climate, periodic famine and natural disasters in many eras throughout different civilizations (Rome, Egypt etc.) obstructed the natural pace of agriculture. The Maya in the Yucatán, the Anasazi in the American Southwest, the Cahokia mound builders outside of St. Louis, the Greenland Norse, the Easter Island statue builders, ancient Mesopotamia in the Fertile Crescent, Great Zimbabwe in Africa, and Angkor Wat in Cambodia have all experienced similar collapses as revealed in recent archaeological discoveries (Weiner, 2018). These civilizations, like many others, perished as a result of a combination of environmental degradation and climatic change, enemy hostility, and diminishing trade with neighbours who were dealing with their own environmental issues. We now can comprehend why societies' declines tend to follow quickly on their peaks since peak population, income, resource consumption, and waste production are all accompanied by peak environmental impact - reaching the point where impact outstrips resources.

Thereafter importance of sustainable agriculture was established by technological innovations in methods like irrigation, ploughing technique, soil preparation and manure application etc. Internal redistribution and external business tend to lessen the influence of local environmental variations on the whole community in agrarian cultures. As a result, the tight spatial linkage between socio-political form, demography, and other factors that shows how cultural fundamental features are linked to the environment is obliterated. However, the pattern of economic and political life, continues to be profoundly influenced by the environment via increasingly complicated causal pathways. Many agrarian cultures, such as those in Indonesia, China, India, Peru and many more, have retained much of their technology and culture. The Homeric tales and the Old Testament are good written records of transition from simpler societies to states in Western and Eastern Asian agrarian societies. Agrarian producers exploited a wide diversity of seasons, soils, and topography as agrarian technology was applied historically from Spain and North Africa in a more or less huge spectrum all across the temperate and subtropical region of Asia to Pacific. Strong inclines always delivered the best culture core concept because they tend to control the effects of history by permitting long-continued contact and diffusion of culture between neighbouring peoples (Montgomery, 2012).

Rapid population growth, opulent lifestyles, and increased demand for resources and services as a result of economic expansion are having a detrimental effect on ecosystem resources. Indiscriminate and inefficient use of natural resource is causing widespread environmental degradation, loss of ecosystem services, generation of excessive waste and additional financial burdens. Climate change and catastrophes, as well as increased vulnerability as a result of unplanned expansion and urban migration, will continue to inflict economic costs that could negate development gains, raise poverty and unfairness, and jeopardize water and food security. While average lifespan is improving, there are many more health risks, which vary by gender and age, as well as expenses associated with widespread environmental pollution, hazardous chemicals and heavy metals, and the appearance and spread of vector-borne illnesses. There has been increasing policy intervention to cope with existing issues but the policy gap is widening due to ineffective policy implementation, a poor scientific base for policy formulation, and rapidly emerging environmental issues. Most of the region's land degradation has worsened, resulting in the displacement of indigenous peoples, loss of biodiversity, and a drop in essential forest products (Singh & Singh, 2017). In terms of soil water content and groundwater recharge, land degradation has significant implications for water supplies (Owuor et al., 2016). Reforestation activities have increased the total forest acreage in various parts of Asia since 1990, yet there are considerable sub-regional variances. Meanwhile, croplands and urban expansion continue to eat away at wilderness, natural forest systems, mangroves, and other natural systems. Natural resources and ecosystems are part of the resource base for sustainable development and a source of livelihoods of rural people. Rapid economic expansion and population explosion, along with rising agricultural and industrial operations and irresponsible performance metrics, have serious consequences for natural ecosystems and human well-being. The Asia-Pacific encompasses 13 of the world's 35 biodiversity hotspots, including the highest number of endemic species, although these ecosystems are endangered by anthropogenic activities (Marchese, 2015). Unfortunately, deforestation in the tropical rain forest zone has a particular threat to species survival. In many regions, plantations are expanding, but they often lack species diversity and do not recover damaged forest ecosystem services. In addition to forests, the products and services provided by marine ecosystems are critical to the region's societal well-being and prosperity. Healthy oceans and coasts are essential not only for food security, but also for alleviating poverty, long-term economic progress, cultural preservation, and pecuniary intermediation. Anthropogenic commotion is undermining many ecosystems'ability to continue to ecosystem services. 60% of world's total population is denizen of Asia and the Pacific which comprised of only 30% of the world's terrestrial expanse. About a half of these populations is facing meagre and 188% augmentation of bio-capacity is the prerequisite to sustain these populaces (Wiedmann & Barrett, 2010).

Rapid economic growth, technological advancements, infrastructure programmes, and population growth and mobility in the region have all contributed to the expansion of agricultural land areas. In some developed countries, such as Australia, New Zealand, and the Republic of Korea, as well as some developing countries, such as Fiji, Indonesia, Malaysia, Papua New Guinea, Samoa, Vanuatu, and Vietnam, agricultural land expansion has been negligible or even decreased in the last decade. Bangladesh, Bhutan, China, India, Pakistan, the Philippines, and Thailand, on the other hand, experienced an increase. In Asia, practices like shifting cultivation continue to have an impact on forest resources. Land tenure in the region requires immediate attention because, despite on-going land reforms, little significant progress has been made, and stalled improvements in land tenure will have a negative impact on land productivity. The demand for agricultural land may increase in the future, with estimates indicating that expanding agricultural land could contribute to a 20% increase in required food production by 2050. However, human activity is reducing many marine ecosystems'ability to continue to provide these services. The massive population of Asia and the Pacific lives on limited resources; in 2011, the region's per person fish consumption footprint was 94% of the global

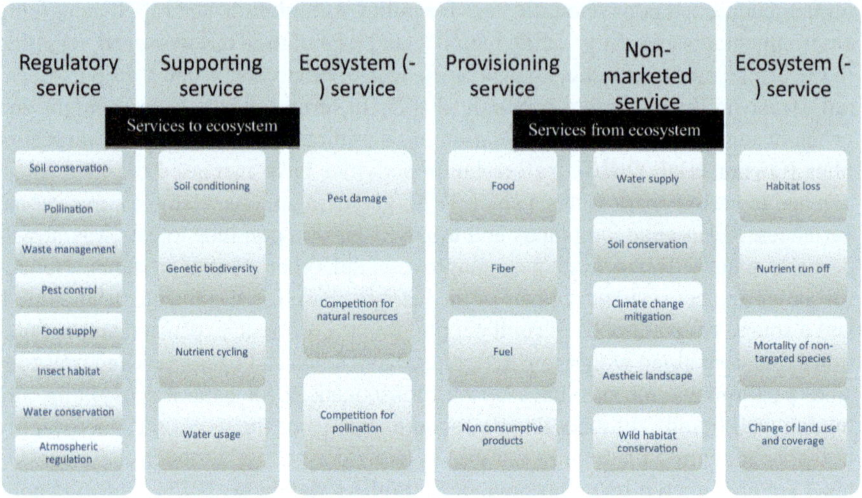

Fig. 1.1 Agriculture ecosystem services: The trade off

average, with cropland footprints of 80%, forest footprints of 63%, grazing footprints of 31%, and built-up footprints of 112%, all resulting in ecosystem service degradation (UNEP, 2016; Guerry et al., 2015) (Fig. 1.1).

1.2 Ecosystem Services and Socioeconomic Identities of Communities

Nature's goods and services are the environmental benefits that result from the physical, chemical, and biological functions of healthy ecosystems. They include market goods produced by ecosystems (for example, food, fibre, fuel, fresh water, genetic resources, and so on), as well as benefits from ecosystem processes (for example, nutrient cycling, climate regulation, water purification, waste treatment, pollination, and so on). Anthropogenic antagonistic activities such as the decline of wildlife habitat and biodiversity can impede ecosystem services. There are several natural capitals from where people able to obtain these services uninterruptedly. Amongst various capital assets like manufactured capital (buildings and machines), human capital (knowledge, skills, experience, and health), social capital (relationships and institutions), and financial capital (monetary wealth), natural capital represents living and non-living components of ecosystems irrespective of anthropogenic creation. It always provides goods and services worth for human. Moreover, the synergies between these capitals yield better results (Fig. 1.2).

Providers affect the generation of ecosystem services and beneficiaries get benefits from the same. Consumers assess the cost benefit ratio of the service based to

Fig. 1.2 Synergies amongst various capitals for provisioning of food

existing policies, distributional channel and involvement of other actors across affected parties. Often policies influence the set of incentives for the private and public use and provision of ecosystem services. Effective supervision and governance can be achieved by the motivation pattern and institutional networking. Payments for ecosystem services (PES) can incentivize potential ecosystem service providers by allowing them to be compensated for taking action, gaining access to a service, or maintaining it. In the same way, rights-based fishery management can encourage fishermen to be better stewards of the ecosystems that generate the fish they collect.

Natural capital that has improved resilience has a stronger potential to adapt to change, continue to deliver ecosystem services, adapt and evolve in advantageous ways. This ability of social-ecological systems to maintain natural capital and ecosystem services in the face of disruption and on-going change is more likely to support development paths in changing contexts distinguished by unpredictability and surprise. Adopting a more integrated dynamic systems approach to comprehending complex social-ecological systems might improve robust solutions that yield desirable outcomes for people and nature under a wide range of potential futures.

Such thinking encourages a broader consideration of possible future outcomes and emphasises adaptive governance. Over the last decade, significant progress has been made in upgrading these principles. Natural capital and ecosystem services are now routinely discussed by influential actors in the public and corporate sectors, scientific research has improved greatly, and new organisations are forming. The fundamental inequalities in economic systems that lead to undervaluing natural capital stewardship have essentially remained constant. To achieve large-scale changes in policies, practices, and investments, it is vital to integrate natural capital and ecosystem services into a broader decision-making context. Such factors are crucial not only to natural resource and conservation decisions, but also to health, agriculture, energy, water security, infrastructure, urban growth, finance, and national security, realms that go far beyond traditional conservation (Figs. 1.3, 1.4 and 1.5).

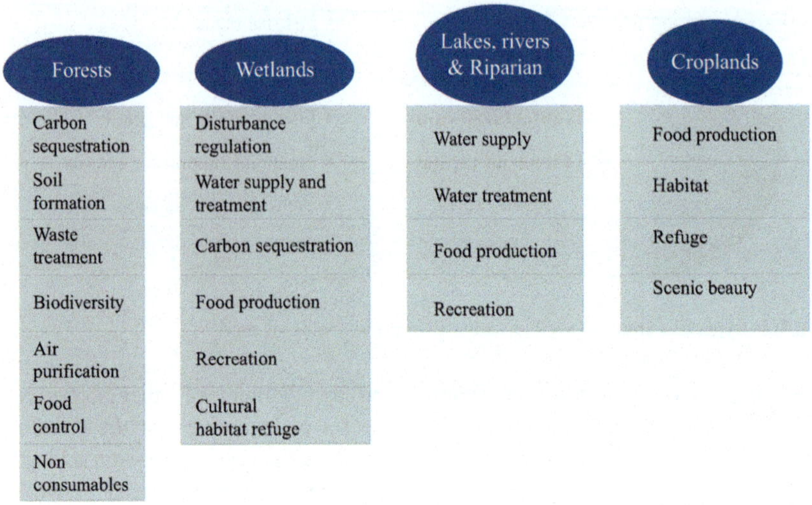

Fig. 1.3 Nature's goods and services derived from natural capital

Fig. 1.4 Ecosystem services

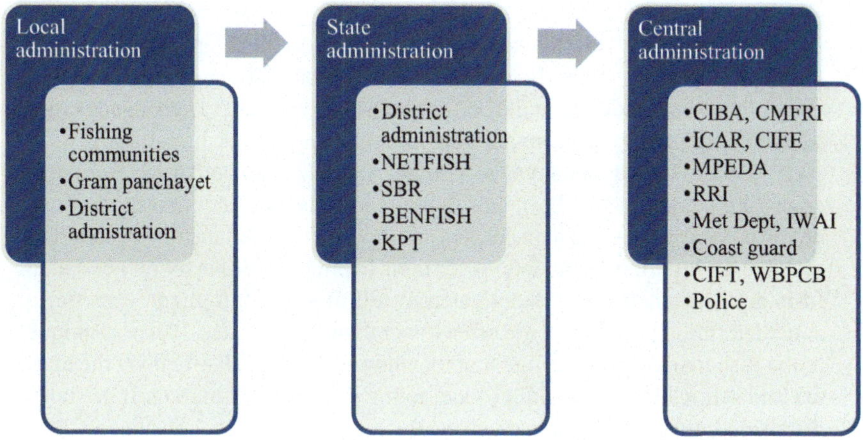

Fig. 1.5 Hierarchical administration for institutional networking and policy recommendation

1.2.1 Case Study 1

'Macher Raja Ilish' (The King of Fishes)

The Indian shad (*Tenualosailisha*) or Hilsa/ ilish known as the Bengal delta's (West Bengal, India) queen/king fish or Bangladesh's national fish. It is socialised with Bengali customs and traditions, and 'Jodailish' is costumed as the bride and groom in wedding ceremonies. From the month of Ashwin (October) through the month of Phalgun (February), fish consumption is discouraged. This coincides with the fish's peak spawning season. As part of the 'Saraswati Puja'/BasantPanchami rite, 'JodaIlish' enters the dwelling once more. Bengalis dislike the taste of khoka/jatkailish. Conservation ethics has thus become an integral component of Bengali culture and ritual.

Bangladesh and India are the two largest fish consumers in South Asia. While Bangladesh accounts for more than 60% of the global fish yield, India accounts for only 15–20%. The Farakka and other barrages harmed spawning, nurseries, and feeding movements, as well as increasing fishing pressure in the upstream river. As a result, ilish and other fish landings in the Ganga-Padma, Brahmaputra-Jamuna, and Meghna Rivers have decreased dramatically in recent years. While ilish has a near-iconic reputation among Bengalis in both India and Bangladesh, little is known about the communities involved in bringing the fish to the cuisine. About 2.5 million people rely on the Ilish fisheries for work and money. Although the species is listed in the IUCN Red List as being of 'least concern', massive extraction has put the livelihood stability of fishermen in both Bengals in jeopardy. The 'bioright' concept focuses

(continued)

on a nature-based solution and a community-oriented approach synergy towards human and natural capital. In Bangladesh, fishing communities are generally considered to be at the bottom of the social ladder. There are descriptions of fisher communities and their lifestyles in popular literature as well, such as ManikBandyopadhyay's 'Padma Nadir Majhi' and A.M. Barman's 'TitasEkti Nadir Naam'. Barman's depiction of the Malo community in great detail is fascinating. Ilish fishermen are completely reliant on the aratdar/ mohajan (who offers loan/ dadon) for their livelihood. Fishers who get dadon from aratdars/ mohajans are obligated to sell their catch to them, sometimes at predetermined prices that are often lower than market rates. Paikar is unique to the ilish marketing system in that they buy fish from fishermen via the aratdar and sell (export) their full product to the international market. It has been discovered that the aratdar receives the majority of the consumer's taka (money) (about 50%), while fishers earn substantially less, depending on the number of middlemen.

In a never-ending cycle of borrowing and lending, the fishing community and the aratdar are linked. There is a lot of room to learn about ilish fishing communities and their livelihoods. Diversification of livelihoods has resulted from lesser capture in rivers, as well as in the estuary and marine zones. Most inland fishermen only engage in seasonal ilish fishing for three months or less. Either, they work as casual labour, migrate to Kerala or Tamil Nadu to work in building projects, work on aquaculture farms in Andhra Pradesh, Gujarat, or Uttar Pradesh, or they set up local food carts selling snacks, among other things, during the remaining nine months. There is also a small percentage of fishermen who work as casual farm labourers. Some people work in agriculture on char lands near the village. In the marine, estuarine, and inland zones, almost all fishermen are landless. The overall capture of fish in the Ganga River system has declined, and the collapse of open water inland fisheries can be seen in the upstream parts of the Ganga between Hooghly, Nadia, and Burdwan Districts. The situation is similar in the marine and estuarine zones. The level of livelihood diversification is confined to migration to urban centres in West Bengal, Kerala, Maharashtra, and Tamil Nadu as casual and/ or construction labour. The 60-day common coastal ban between April and June, as well as trawling and mechanisation of boats in the marine and estuary zones, have increased catches in the post-ban season. Fishermen in the maritime and estuarine zones have stated that the next three months (July, August, and September) are the busiest months for fishing, particularly ilish, with earnings lasting at least six months, if not longer.

However, there are two significant advances in the ilish that must be addressed. Following a direction from the Indian government to ilish-producing states in 2011, the state of West Bengal amended its marine and inland fisheries acts in 2013 to ensure ilish conservation. Regulations on mesh size, the use of monofilament gill nets for ilish fishing, and a fishing ban between

(continued)

September and October, according to the lunar calendar, are among them. The 2013 amendment also selected several Hooghly River segments as ilish breeding/spawning zones and designated them as proposed ilish sanctuaries. The Hilsa Conservation and Research Centre was also established as the primary coordinating agency for ilish conservation and livelihoods. It should be noted, however, that the coastal ban between 15 April and 14 June is now well-established, and one can only hope that, in exchange for higher returns in subsequent years, the fisheries federations will be organized/ convinced to adhere to the 11-day (or 15-day) fishing ban during the peak ilish spawning season between September and October. Overfishing, as well as other issues such as salinity, turbidity, and pollution, have an impact on the ilish migration. It is critical to focus on deeper institutional coordination between multiple departments, especially local government institutions and ministries that have a stronger role in river management or being stakeholders, in order to combat declining ilish stocks. In the context of ilish, fisheries co-management in West Bengal could be a suitable place to start. It's time to examine and better understand awareness, outreach, and other livelihood possibilities. The term 'complementary livelihoods', is used to describe occupations that are similar to fishing in some way. Positive policy responses are an opportunity that must be capitalised on in order to advance the aim of balancing conservation and livelihoods. Despite the fact that fisheries departments in both nations are responsible for the fish and the livelihoods, there is a complicated network of institutions that must be involved to maintain the initiatives and long-term viability (Mandal et al. 2018).

1.2.2 Case Study 2

'Pantabhasani Baan' (The Superfluous Flood)

'Pantabhasani Baan' means the excellence level of high tide during month of October every year and it is allied with some auspicious day of 'Mahalaya' before 'Durga Puja' the grandest festival of West Bengal. The tidal amplitude remains very high so that embankment breaching becomes almost certain and the saline water inflows into the villages and huts. The villagers suffer a miserable condition and face heavy loss of livestock, food, drinking water, belongings etc. The earthen embankments (around 3500 km^2) around Indian Sundarban are almost 150 years old and are being weakened everyday by the swirling currents that scour at their bases and by tidal surges coupled with strong winds and cyclones. During the rainy season, during 'Kalbaisakhi' and even throughout the year embankment breaching takes place. As a result, the

(continued)

local communities suffer a lot. Due to embankment breaching saline water enters the agricultural land and pond. So, agriculture and fish farms are badly affected. Even mud-built houses collapsed and seasonal flooding as well as tidal flooding also occurs. This is a serious problem of the area. The villagers of Gobardhanpur village of 'Buraburir Tot' in G plot of Patharpratima block, South 24 Parganas, of Indian Sundarban, every year face same kind of meagre condition even worse than any other villages of Indian Sundarban. As it is situated at the southernmost tip of land mass of Indian Sundarban and face the Bay of Bengal. The villagers are basically engaged in capture fishery as their seasonal livelihood. Some agricultural lands are there inside the village but inundate every time as a consequence of embankment breaching. The southernmost barren mudflat never favoured natural mangrove regeneration as it was used as harbouring site of fishing boat (non-mechanized).

Women community of that village was engaged in meen (juvenile shrimp seed) collection. They realized the non-performance status of meen collection profession but had no alternative choice to sustain their livelihood. Biorights concept proved to be a boon for this village. The women community was agreed for capacity building in backyard poultry farming and climate change resilient aquaculture. They encouraged their partner for the same and the entire community was engaged in backyard poultry farming and aquaculture in homestead pond. The state administration helps the community with their existing schemes and supports. To combat embankment breaching the community received funds from different NGOs. They afforested selected species of mangrove on the barren mudflat. These mangroves are serving as primary protection shield against the 'Pantabhasanibaan' (Baanerjolghorerpantabhasiyeniye jay- the flood water insert into the house and take away all material including very cheap watered rice of previous day. The villagers earn their livelihood from this estuarine region, so they are the prime actors to conserve the environment. Their mangrove restoration approach was nothing but paying back or nurturing the nature for utilising its goods and services.

1.3 Ethnobiology and Natural Rights of Communities

Socioeconomic and cultural identities of marginal communities are strongly integrated with the goods and services of the ecosystems they live in. Its, therefore, imperative that any conservation initiative related to land and other natural resources of that ecosystem will have direct linkages with their rights and liberties. In order to ensure climate transparency and accountability in enhancing long term opportunities for conservation of habitat biodiversity and ecosystem services the initiatives and actions, be it implementation or policy planning, must respect the community rights and endure the obligations that the same is being protected and promoted as well. In the course of protecting ecosystem integrity, conserve habitat and biodiversity or sustainably intensify the ecosystem services, there would be some obvious impacts on the rights and liberties of people, as it affects their access to land and natural resources. Vulnerable and marginal communities in poverty traps are frequently found to be living in and around resource rich ecosystems and high-biodiversity areas as they depend mostly on the goods and services of the ecosystem to fulfil their needs and requirements. Nevertheless, in many places communities themselves preserve and protect the ecosystems too, to become self-reliant. Such reciprocal interdependence is often found in wetland areas, coastal mangrove habitats and highland forests in mountains of global south. However, in the changing regimes of climate and parallel policy formulations in abating the impacts of change, there has been issues and concerns about the community rights triggering conflicts that undermines the conservation objectives. We will discuss such case studies in the following chapters for furthering this discussion. It is important for all implementing agencies and lead development organizations to undertake pre-appraisals for assessing the potential impacts of the planned conservation actions on the community rights and livelihoods of the people therein. Ethically, a free pre-intimated consent of the community members must be facilitated to avoid any such adverse impact and pledge for equitable partnership.

Global frameworks for conservational thinking and practice have duly recognized these ethical and pragmatic concerns about the well-being of communities and their rights providing a rationale for conflict resolution. Several interventions and policies like integrated conservation and development projects (ICDPs) in 1980s and 1990s, along with recent approaches like building on lessons about the limitations of ICDPs such as co-management, community-based natural resource management (CBNRM) and recognition of indigenous and community conserved areas (ICCAs) are few to mention. Climate initiatives like Ecosystem-based Adaptations (EbA) and Nature-based Solutions (NbS) are also in tune with these.

1.3.1 Ethnobiological Significance of Community Rights

Indigenous and local communities'collectively held knowledge, innovations, and practises that are inextricably linked to traditional resources and territories, local economies, genetic diversity, species and ecosystems, cultural and spiritual values, and customary laws shape the socio-ecological context of communities. When it comes to conserving traditional knowledge and biocultural heritage, community rights are always important. Traditional knowledge is the result of generations of development by indigenous and local communities, who have de facto ownership of it. The moral imperative of ensuring these rights is upheld and other key reasons include:

- To implement CBD Articles 8 (j) and 10 (c): The Convention on Biodiversity article 8(j) recognises the importance of 'the knowledge, innovations and practices of indigenous and local communities' for the conservation and sustainable use of biodiversity. It requires Parties to 'respect, preserve and maintain' them, 'promote their wider application with the approval and involvement' of communities and 'encourage equitable sharing of benefits from their use'. The rights of communities over their knowledge are implicit in these three clauses, and need to be recognised for their effective implementation. Furthermore, CBD article 10 (c) requires Parties to 'protect and encourage customary use' – this depends on continued use of traditional knowledge by communities, and hence on protecting their rights over it.

- To create conservation incentives: The CBD requires Parties to develop biodiversity conservation incentives. The Plan of Implementation of the 2002 World Summit on Sustainable Development emphasised the importance of local benefits and incentives. The value of local benefits and incentives in supporting conservation has been demonstrated in numerous studies (e.g. the Millennium Ecosystem Assessment, 2005). To enable communities to create benefits, secure rights to traditional knowledge and related resources are required, and these rights can operate as incentives for long-term management.
- Traditional knowledge is used in adaptive management systems that constantly adapt local bio-resources, knowledge, and practises to environmental change in order to adapt to climate change. Such adaptive solutions are critical for farmers worldwide, not just TK-holders, to adapt to climate change. Traditional knowledge that has been documented is no longer changing to meet new challenges. However, as community land is taken over (for industrial farming, forestry concessions, biofuels, conservation, and so on), and cultural values and customary rules are weakened, traditional knowledge systems in situ are rapidly disappearing.
- For adaptation to climate change: traditional knowledge forms part of adaptive management systems which continually adapt local bio-resources, knowledge and practices to environmental change. Such adaptive systems are vital for adaptation to climate change, not only by TK-holders but by farmers worldwide. Documented traditional knowledge is no longer evolving to address new challenges. But traditional knowledge systems in situ are fast disappearing as community land is taken over (for industrial farming, forestry concessions, biofuels, conservation etc.), and as cultural values and customary laws are eroded. This calls for protection of community rights over their bio-cultural heritage as a whole – traditional knowledge, bio-resources, culture and land.

Indigenous and local communities have their own customary laws and values which set out their rights and responsibilities relating to traditional knowledge. These are designed to ensure traditional knowledge meets community needs and is maintained for future generations. Customary laws evolve to address new challenges and are vital for the way of life of indigenous peoples and their customary use of resources to continue. Yet, as this project found, customary laws and institutions have been weakened and eroded by various factors, including extension of government authorities and natural resource laws, and alienation of community land. Hence, responses are urgently needed which recognise and strengthen customary laws and authorities.

Integration of rights-based approaches with conservation objectives also brought newer dimensions in conservation practice. These elements were 'Grounding in defined standards based on international human rights frameworks' and 'Relationships of accountability between "rights-holders" and "duty bearers", where 'right holders' are the communities and 'duty bearers' are the developmental agents for change management. Relationship accountability defined the boundaries of action and inaction to protect the rights, prompt the responsibilities and promote transparency.

Local knowledge and ecological practises have played a significant role in the study of how people create places and experience the environment. In today's 'ecological era,' this topic has regained relevance. Different processes occur, however, which can appear to be philosophically ambiguous in some ways. Deductive logics and etic techniques are permeable to ethnobiological research, which are heirs of linguistic and cognitive anthropology. The ethnoecological paradigm is politically located and related with the discourse of 'indigenous rights' (CBD, 2010, Frazão-Moreira, 2015).

1.3.2 User Rights and Human Rights: Conflicts and Convergence

Introduction of 'Right-based approach' made a paradigm shift in development sector as UN organs and several nongovernmental international development agencies adopted institutional right based approach demanding recognition of rights as a basis for any partnership – particularly with indigenous peoples, who are working towards their indigenous rights through international and social movements. Thus, indigenous rights became the first and the most robust engagement on rights issues, such as security of land tenure, right to effective outcomes of the partnership and strengthening recognition of community governances in matters that relate to conservation and natural resource management. On the other end, accountability of organizations included both State actors and non-State actors as primary duty-bearers to also share certain human rights responsibilities, even if the law of the land may fail to hold them accountable. In all cases, community rights are now systematically considered with development activities, wherever they are impacted or affected, whether both are in synergies with conservation objectives or have trade-offs and conflicts in between them. While this protects the violation of human rights or community rights of marginal people, their bearings or costs incurred by them directly or indirectly in the course of conservation by forgoing the opportunity to liberally use the goods and services are not protected or compensated as most of it is either non-valued or intangible in nature. Here lies an economic lacuna in equitably engaging the right-holders in the act of conservation. It is presumed that such compensatory packages would incentivize their efforts in conservation and develop both onus of conservation and ownership of nature. In fact, the dimensions and magnitudes of specific human rights issues are necessarily local and context-based and so is the solution to it. Thus, in addressing it with a localized approach, a complex process comprising specifics of local ecology, history, culture, political context and governance system would be obligatory. However, all solutions at the same time must ensure the fundamental standards of human dignity defined in human rights frameworks.

Rights of communities and human rights as well have multiple sources of origin that includes statutory legal systems and customary community laws of indigenous people. This plurality of perception of rights has both challenges and prospects embedded in it. Definitely this makes it difficult to apply universal standards, prescribed in various charters, but the diversity in it creates opportunities for designing localized adaptive solution. At the same time, when the customary laws are recognized and the processes and institutions in the locale are regarded, it enables an equal reciprocal partnership. It's true that in a right-based approach, activities of interventions are often entangled between the inter-polarity or intra-polarity of power among right-holders and duty-bearers pertaining to decision making and governance. This either enables or disables the activity links with the environment that is shaping the conservation objectives but more often it infuses scopes for better negotiations and advantages to leverage the outcomes. It is obvious that rights-based conservation will often require appropriately addressing complex issues of governance and empowerment of vulnerable rights-holders, failing which superficial engagements will be an immediate challenge to sustainability. This is bestowed more on the non-state actors rather than state actors playing the primary duty-bearer's role. State actors fail to perform adequately owing to policy gaps in national legislations and international principles, while poor governance, lack of resources and capacities may also be some other reasons pertaining to their incapability or inaction. It must also be recognized here that civil societies or NGOs and other non-state actors also have their own limitations and they can neither fulfil nor protect the rights of the communities, unless they are empowered by the policy instruments. It may even so happen that they (non-state actors, CSOs and NGOs) have to withdraw from the community, losing potential chances of transformations, just because of compliances and obligations to the rights of the communities. Therefore, in this restricted window of operation, self-regulation of non-state actors must be ensuring the social responsibility of recognizing the community rights and as well defuse conflicts to enable the conservation objectives. Additionally, they must also promote the right-based approach at community-ecosystem interface, if the existing policies, frameworks or local customary laws do not establish it in it's true standards and principles. Similar plurality of perception exists about the non-state actors, especially CSOs and NGOs in the eyes of communities and they are considered quite powerful and resourceful as well, but the confusion remains in their intensions of actions, whether they are in favour of the communities or on the other side. Herein too, the fiscal model of compensating their forgone opportunities in contributing to conservation activities can assure a win-win situation furthering the partnership to sustainability.

Procedural rights are those which addressed by accessing the process of asserting them through a legal or customary framework. The awareness of communities about their procedural rights enables them to negotiate about their entitlements and realizations of substantiative rights. Establishing the procedural rights provides a space to make effective claims in systems of mutual accountability. Some common procedural rights are:

1. Right to information about the planned developmental and conservational activities, its implications on natural resources and its usage, threats and benefits from the local environment etc.
2. Right to participation enables the community beneficiaries to actively participate in the decision-making system, towards building an equitable and meaningful partnership.
3. Right to Justice (access and redressal) allows the right-holders to raise concern about the denial and/or disregard of their rights and seek access to redressal for the needful.
4. Substantive rights include the right to life, lifestyle and a dignified livelihood in a healthy environment. This is an important section that applies to conservation efforts as it recognizes the sustainable availability of ecosystem goods and services for a dignified livelihood within the realm of healthy environment. Restrictions on physical and economic access to these resources, even for the cause of conservation, therefore, becomes an infringement to this procedural right.
5. Right to health also substantially include environment and recognizes an unpolluted and sustainable ecosystem for health and well-being. In the post pandemic period, issues and concerns pertaining to right to health are strongly advocated.
6. Right to adequate standard of living is mostly governed by the sustainable economic and physical access to and availability of safe and culturally appropriate food of sufficient quality and quantity. There are a number of relative terms in the definition e.g., 'adequate', 'culturally appropriate' etc. that increases the plurality of the right to its advantage but appears to be a bottleneck before the conservation targets. UNHRC recognizes right to water and sanitation as an integral part to this right as it pertains to standard of living. In this regard adequacy and budgeted usage of water may provoke conflicts, unless explained in the right context.
7. Right to development acquires equitable and participatory partnership in economic, societal and political development reinforced by infrastructural facilities. Naturally the implications are fervidly debated in the conservation and climate milieu.
8. Cultural rights of indigenous communities permit them to practice and pursue their traditional activities and rituals that establish their cultural identity. This has special implications on biodiversity loss and conservation as well. It widely varies among communities and socio-ecological landscapes that how the communities manage and mitigate their cultural dependence on biological species in the locale.

9. Right to work help ensure a healthy work environment for a dignified livelihood. Particular issues related to labour laws, employment and income generating activities linked to conservation projects, commodity certification standards etc. may be associated with Right to Work. In post pandemic lockdown period, reverse migration of casual labourers has raised trivial issues related to livelihood securities in the ambit of this right.
10. Right to property is a statutory right to possess a property unless the existence, value, or use, of that property is enjoyed at the expense of others loss or deprivation and without due consideration of and informed consultations with others who also have similar rights in the property. Especially, in case of natural enclaves those are valued for community utility and usage and conservation can impede such rights and therefore cause conflicts.
11. Right to self-determination and use of natural wealth and resources allows people (especially from indigenous communities) not to be deprived of means of subsistence. Though these definitions have implications related to sustainable resource use through responsible consumption and production, wherein conservation objectives may appear highly conflicting. However, this is also true that sustainability and conservation is in the central tenet of indigenous culture and practice. International conventions on indigenous peoples recognize their rights to traditional lands, territories and resources.

1.3.3 Societal Paradigm for Indigenous Users of Nature's Goods and Services

Risk of infringement in the basic rights of marginal communities is a major concern as it makes them more vulnerable. Thus, the universality of human rights always protects them by preventing infringements that can be very well from developmental activities, urbanization and economic growth. The socio-economic challenges of the marginal communities make them lean over the natural resources for livelihood and wellbeing. Whenever there is a conflict with the outside interests vested on the over exploitation of these high value resources like mineral, forests or fisheries, their vulnerability is enhanced. This is more because of their relative lack of power and political weaknesses. The communities fail to defend their rights and succumb to exploitation. Even within their own communities, women and indigenous people experience greater vulnerability owing to socio-political marginalization. Especially for the indigenous set of people, conflicts are furthered if their local institutions and customary claims are not recognized by the statutory law of the land. There are several such case studies, wherein self-conflict between procedural rights too can create a breach rendering the indigenous people more vulnerable to external thrusts. The role of a robust structural framework for protecting their substantive rights with due consideration of such possible areas of conflicts and doubts becomes crucial, as laggings therein may be detrimental to their sustainability and even survival. Recent times international rights instruments, e.g., ILO169 and UNDRIP, specifically addressed to indigenous people's individual rights such as access to resources for food, health and shelter and collective rights, such as to territory and land tenures etc. However, the framework instruments vary widely across the globe and unswervingly depend on the socioeconomic profile of the ecoregion as a whole.

Traditional ecological knowledge of the indigenous people, their control measures and management of lands and resources, and local indigenous institutions for self-governance substantially contributes to conservation. This enables the 'Rights to traditional lands, territories and resources' alongside "right to the conservation and protection of the environment and the productive capacity of their territories and resources." These instruments and as well correct interpretations of the procedural rights makes space for certain logical and committed areas of considerations in engaging the marginal communities or indigenous people in conservation partnerships. Free, pre-intimated information sharing is one such area that empowers the people's participation, whereas rights to equitable benefit-sharing can enable their partnerships in determining the development happening through an intervention or help them identify the priorities and strategies for conservation of their lands, territories and resources and to benefit equitably from conservation and sustainable use of such areas and resources.

Gender just participation in decision-making and women's control over natural assets such as land is equally important as women often suffer from discrimination that can limit their distinctive roles in natural resource management. In addressing this inequity Convention on the Elimination of All Forms of Discrimination against Women (CEDAW), was adopted in 1979 and still remains to be the main international instrument addressing the equal rights of men and women. It ensures that they both on a basis of equality, participate and benefit from rural development, access economic opportunities and leverage from land and agrarian reforms as well.

1.3.4 Key Issues at the Intersection of Conservation and Human Rights

LNOB Principle: Conservation contexts in which the lines between community rights and their protection or fulfilment can often not be strictly drawn, proactive investments in building capacities is required to ensure participation, leaving no-one behind.

Doing-no-harm: This requires developing a deep understanding of cultural, socio-economic, political or historical contexts to understand the harm in breaking the links between conservation and community rights.

Community governance: Issues pertaining to rights of either people or community that arise in conservation are largely a function either enabling or disabling the broader environment in the context of place-based governance in which conservation is being carried out. This implies that in a constrained environment meaningful progress in rights-based conservation is unlikely for policy engagement and improved governance.

Strategic Partnership: Strategic partnerships at grassroots can help to augment the influence on policy or bureaucratic systems in fulfilling the rights of the communities, which otherwise cannot be guaranteed through any obligatory protection that the government can extend, as civil societies and non-state actors have their own limitations of stimulating the system.

Legal Pluralism: Tenure security of a land or property can get trapped in the gap of customary and statutory laws, resulting in conflicts between right holders and duty bearers. If the tenure security is weak, any traditionally owned land may get termed as a vested land to be owned by the state. Conflicts owing to such overlaps in legal paraphernalia not only denies their substantive rights but also make the right holders vulnerable to ensuing changes. To evade this ILO 169 and UNDRIP have specifically recognized collective rights of indigenous peoples to traditional lands, territories and resources. Weak tenure security diminishes the rights of local communities is also a challenge before sustainable conservation and restoration efforts.

Cultural rights and bio-cultural identity: There are often strong positive synergies between rights of communities to their cultures and biodiversity conservation objectives. Healthy natural systems and species diversity beholds a number of traditional practices and community rituals that propounds the bio-cultural identity of a community or indigenous group. Recognizing their cultural rights, therefore, conserves the species and natural systems.

1.3.5 Equity in Benefit-Sharing

For many rural people, a healthy environment is a primary foundation for development and generation of economic benefits. Thus, there are strong motivations for rural people to use natural resources in ways that maintain these key "assets" for the future. At the same time, even where communities have strong ties to and dependence on natural resources, it is unrealistic to expect that they will forego opportunities for poverty reduction, especially to meet basic needs. There are inherent tensions in the Brundtland Commission's definition of sustainable development – "to meet the needs of the present without compromising the ability of future generations to meet their own needs" – as conserving resources for future generations often involves limitations or modifications on present use and/or limitations on future development options. Efforts to resolve this tension through support for low impact activities have sometimes been criticized for locking in subsistence-level livelihoods. One way that sustainable development and conservation may be made more compatible is through the generation and equitable sharing of increased benefits from natural resources, through "green economy" and alternative economic growth paths, or (often in more limited ways) from additional income streams linked to protected areas or payments for environmental services. Questions of equitable benefit-sharing also often arise in the context of broader development activities carried out on community lands, by community members themselves, or through agreements with outside actors (commercial forestry, agriculture, energy development, mining, etc.). Widening the concept of development benefits beyond material and commercial benefits can also be a way of reconciling conflicts between conservation and sustainable development. Guiding Principles for respecting sustainable development rights, and further contributing to wellbeing through, inter alia, benefit generation and equitable benefit sharing, include the following: In all cases, do not exacerbate poverty or make vulnerable people worse off (do no harm); Implement "constructive programs that benefit local communities, such as capacity training that can provide income alternatives (especially where use reductions are necessary) or assistance in diversifying management capacities and implementing sustainable use methods" (from Addis Ababa Principles). This can include facilitating access to financing and markets, and other supportive actions. Ensure that "an equitable share of the benefits of sustainable use regimes remain with the local people (in the case of foreign investment), considering both monetary and non-monetary benefits" (from Addis Ababa Principles). Ensure that diversified livelihood activities support/do not economically or socially displace the most vulnerable people within communities. Recognize and respect communities' own values and visions of development. Practical approaches to respecting rights in relation to sustainable development and equitable benefit sharing include to: Conduct participatory research to understand the development priorities and plans of indigenous peoples and local communities, and to assess social impacts of conservation activities (including all costs and benefits, and their distribution). Avoid undertaking or supporting conservation initiatives that unnecessarily restrict sustainable resource use, or that undermine equitable benefit/cost sharing. Practical approaches to supporting protection and fulfilment of rights in relation to development and benefit-sharing include: Support rights-holders' capacity and opportunity for enhanced sustainable development and benefit generation and sharing through conservation (e.g., various local and landscape-level natural resource-based community development activities). Support other duty-bearers' (e.g., government, private sector investors, civil society actors) understanding, capacity, and commitment to supporting sustainable community development, e.g., through training, pilot projects, information sharing, and dialogue. Working with appropriate partners, promote and facilitate policy changes that support greater opportunity for sustainable development and benefit sharing (e.g., substantively decentralized natural resources management), and that put in place protections for people's basic rights where restricted access/use is necessary (Frazão-Moreira, 2015).

References

Frazão-Moreira, A. (2015). Ethnobiological research and ethnographic challenges in the "ecological era". *Etnográfica. Revista do Centro em Rede de InvestigaçãoemAntropologia, 19*(3), 605–624. https://doi.org/10.4000/etnografica.4141

Guerry, A. D., Polasky, S., Lubchenco, J., Chaplin-Kramer, R., Daily, G. C., Griffin, R., … Vira, B. (2015). Natural capital and ecosystem services informing decisions: From promise to practice. *Proceedings of the National Academy of Sciences, 112*(24), 7348–7355. https://doi.org/10.1073/pnas.1503751112

Iqbal, S. (2017). Impact of environment on architecture of Mesopotamia with respect to the use of materials. *Tools and Mode of Construction.*. https://ssrn.com/abstract=3055608

Kineman, J. & Anand, D. (2014). *Roots of sustainability in ancient India.* 58th Annual Meeting of the International Society for the Systems Sciences, ISSS 2014.

Marchese, C. (2015). Biodiversity hotspots: A shortcut for a more complicated concept. *Global Ecology and Conservation, 3*, 297–309. https://doi.org/10.1016/j.gecco.2014.12.008

Montgomery, D. R. (2012). *Dirt: The erosion of civilizations.* Univ of California Press.

Nishat, B., Mandal, S., & Pangare, G. (2018). *Conserving ilish, securing livelihoods: Bangladesh-India perspectives.* Academic Foundation in association with The International Water Association.

Owuor, S. O., Butterbach-Bahl, K., Guzha, A. C., Rufino, M. C., Pelster, D. E., Díaz-Pinés, E., & Breuer, L. (2016). Groundwater recharge rates and surface runoff response to land use and land cover changes in semi-arid environments. *Ecological Processes, 5*(1), 1–21. https://doi.org/10.1186/s13717-016-0060-6

Secretariat of the Convention on Biological Diversity. (2010). *Ecosystem goods and services in development planning*. A Good Practice Guide.

Singh, R., & Singh, G. S. (2017). Traditional agriculture: A climate-smart approach for sustainable food production. *Energy, Ecology and Environment, 2*(5), 296–316. https://doi.org/10.1007/s40974-017-0074-7

Swiderska, K. *Protecting Community Rights over Traditional Knowledge. Implications of customary laws and practices.* Key findings and recommendations 2005–2009. International Institute for Environment and Development (IIED), 3 Endsleigh Street, London WC1H 0DD, Website: www.iied.org

UNEP. (2016). *GEO-6 regional assessment for Asia and the pacific*. United Nations Environment Programme. http://www.des.ucdavis.edu/faculty/richerson/BooksOnline/He6-95.pdf

Weiner, M. H. (2018). *The collapse of civilizations*. Paper Published by Belfer Centre for Science & International Affairs, Harvard Kennedy School. https://www.belfercenter.org/sites/default/files/files/publication/TheCollapseofCivilizations.pdf

Wiedmann, T., & Barrett, J. (2010). A review of the ecological footprint indicator—Perceptions and methods. *Sustainability, 2*(6), 1645–1693. https://doi.org/10.3390/su2061645

Zhang, Y., Min, Q., Li, H., He, L., Zhang, C., & Yang, L. (2017). A conservation approach of globally important agricultural heritage systems (GIAHS): Improving traditional agricultural patterns and promoting scale-production. *Sustainability, 9*(2), 295.

Chapter 2
Rights Based Approach in Conservation

Malancha Dey Roy

2.1 Community Rights in Conservation

Conservation of nature and natural resources is inextricably linked to people's rights to secure their livelihoods, live in healthy and productive settings, and live-in dignity. The pursuit of conservation goals can aid in the fulfilment of a variety of basic human rights. Secure rights, such as land tenure and decision-making involvement, can also help to improve environmental stewardship. Conservation initiatives, on the other hand, can have negative consequences if their connections to concerns of human rights and well-being are not fully recognized or handled, and poor rights enforcement can sabotage conservation efforts. Many challenges remain about how to successfully integrate human rights ideas into conservation efforts. Implications for conservation strategy design and execution across a number of programmatic areas (e.g., climate change, protected areas), particularly where there are conflicts between different rights and rights holders, or where conservation aims contradict. Other questions concern the policies, capabilities, and procedures that conservation groups will require to implement and monitor their human rights commitments. The implementation of international human rights frameworks, as well as major rights problems pertinent to conservation, require common understanding. Any conservation method that recognises rural people's deep economic and cultural ties to land and natural resources has the potential to significantly contribute to the achievement of human rights. Responsible acts that are conscious of others' rights, as well as commitments to help respect, defend, and promote those rights, contribute to a climate of openness and accountability that improves long-term conservation potential. Efforts to maintain biodiversity and protect ecological integrity, on the other hand, have an impact on people's access to land and other natural resources, and

M. D. Roy (✉)
South Asian Forum for Environment, Kolkata, West Bengal, India
e-mail: research@safeinch.org

D. Dey, J. Roy (eds.), *Biorights*, Studies in Ecological Economics 7,
https://doi.org/10.1007/978-3-030-91503-2_2

hence their enjoyment of human rights. People living in high-biodiversity areas are frequently among the most vulnerable in a country, and their livelihoods are directly dependent on ecosystem services and access to land and natural resources. Conservation groups, from an ethical stance, have a responsibility to identify and resolve the potential negative effects of conservation initiatives on individuals whose rights or livelihoods may be harmed. Conversely, failing to be explicitly aware of all actors' rights in a given area can lead to or exacerbate conflicts over land, water, and other resources, undermining conservation efforts. Fundamental ethical and pragmatic concerns about human well-being have been part of conservation thinking and practise for decades, providing a rationale for integrated community development projects (ICDPs) in the 1980s and 1990s, and underlying more recent approaches (building on lessons learned from ICDPs) such as co-management and community-based natural resource management (CBNRM) and recognition of indigenous and community conserved areas (ICCAs).

2.2 Environmental Rights

Environmental rights are separate from, but strongly connected to, more commonly recognised human rights. They are often regarded as a human right to a clean, healthy, and safe environment. 'Man has the fundamental right to freedom, equality, and adequate conditions in an environment of a quality that permits a life of dignity and wellbeing, and he bears a solemn responsibility to protect and improve the environment for present and future generations,' according to the Stockholm Declaration (Declaration of the United Nations Conference on the Human Environment) (1972). (Principle 1). Despite not being approved, the Draft Declaration of Principles on Human Rights and the Environment (1994) represents a broad attempt to make environmental rights more explicit in human rights. It asserts that everyone has the right to a healthy environment that fairly satisfies current generations' needs without jeopardising future generations' ability to satisfy their own. It also emphasises that everyone has a responsibility to safeguard and maintain the environment. The UN General Assembly and the Human Rights Council both adopted explicit wording on the right to water as a human right in 2010, including clean drinking water and sanitation. The African Charter on Human and Peoples' Rights, the 'Protocol of San Salvador' (Additional Protocol to the American Convention on Human Rights in the field of Economic, Social, and Cultural Rights (1998), and the Revised Arab Charter on Human Rights all acknowledge environmental rights (2004). Nearly 115 country constitutions currently incorporate references to rights to a safe, healthy, or environmentally balanced environment. Environmental rights, including non-human rights of the environment and natural systems, have also been addressed in indigenous peoples' pronouncements, such as the 'draft Universal Declaration of the Rights of Mother Earth'. The environment in which we live is fundamental to all of us. The full enjoyment of a wide variety of human rights, including the rights to life, health, food, water, and sanitation, requires a safe, clean, healthy, and sustainable environment. We won't be able to achieve our goals unless we live in a healthy atmosphere. We may not have access to even the most basic human dignity criteria. At the same time, defending human rights contributes to environmental protection. People may contribute to ensure that decisions that impact them respect their desire for a sustainable environment by learning about and participating in the decisions that affect them. Political rights, such as rights for indigenous peoples and other collectivities, the right to information and involvement in decision-making, freedom of opinion and expression, and the right to prevent undesirable developments, are all included in our understanding of environmental rights. We also believe in the right to seek justice for rights violations, including rights for climate refugees and anyone displaced by environmental degradation, as well as the right to demand ecological debt and environmental justice. Many of these rights, especially political rights, are well-established and codified in numerous treaties and agreements. We can appreciate the ongoing efforts of communities and indigenous peoples around the world for the formation of some of these rights, as well as the acceptance of others that are not yet formally defined. Other 'new' rights, such as rights for climate refugees, have emerged in recent years as a result of the rapid pace of economic globalization and the resulting environmental and social consequences. Others, such as the entitled to demand ecological debt, have evolved as a result of years of campaigning by Friends of the Earth and others to acknowledge the effects of northern resource depletion and environmental damage in southern countries. All of these rights are equally vital, and they all depend on one another. Environmental rights are human rights because the quality of and access to the surrounding environment, as well as the recognition of people's rights to knowledge, participation, security, and restitution, all affect people's livelihoods, health, and sometimes even survival.

2.3 Rights of Indigenous Peoples

Indigenous peoples' rights have been a focus of recent advances in international human rights legislation, in acknowledgment of indigenous peoples' unique status as separate cultures, frequently in non-dominant positions in national society, and with strong links to ancestral territory. Convention 169 Concerning Indigenous and Tribal Peoples in Independent Countries (International Labour Organization (ILO) 169) (adopted 1989, entry into force 1991), which was one of the first instruments to recognise indigenous peoples' relationships and rights to lands and natural resources that they traditionally occupied. The Convention forbids the removal of peoples except in exceptional circumstances and with free and informed consent; and the Universal Declaration on the Rights of Indigenous Peoples (UNDRIP 2007), which, while non-binding, 'represents one of the most important developments in the promotion and protection of indigenous peoples' basic rights and fundamental freedoms. It covers self-determination, land and resource tenure, free, prior, and informed consent in respect to their lands and territories, as well as the enjoyment of cultural and customary institutions. Indigenous peoples' unique situation, which includes the preservation of customary institutions and ties to their traditional lands, has resulted in provisions on indigenous peoples in core human rights treaties, as well as – in more recent decades– international rights instruments specifically addressed to indigenous peoples (particularly ILO 169 and UNDRIP). Indigenous peoples' rights include collective or group rights, which are connected to their individual rights, because collective rights (such as territorial rights) are frequently necessary for individual rights to be realised (such as to food and health). The acknowledgment of indigenous peoples and their rights by states and other actors varies significantly by area; for example, although well progressed in Latin America, indigenous peoples in Africa and other regions continue to battle for recognition. Due to the frequent overlap of high-biodiversity regions with indigenous territories, and the vulnerability of natural resource-dependent customary livelihoods to changes in access or usage, indigenous peoples' rights are frequently particularly essential for conservation and sustainable use of natural resources. Indigenous peoples' traditional ecological knowledge, traditional systems of control, use and management of lands and resources, and traditional institutions for self-governance also contribute substantially to conservation (Springer et al., 2011). Aspects of indigenous rights especially relevant in conservation contexts include (Table 2.1):

Table 2.1 Indigenous rights in conservation context

Rights	Elucidation
Rights to traditional lands, territories and resources	Including the "right to the conservation and protection of the environment and the productive capacity of their lands or territories and resources."
Rights to self-determination and to free, prior, and informed consent (FPIC)	Self-determination is a collective right reflecting indigenous peoples' status as distinct peoples. Universal declaration on the rights of indigenous peoples (UNDRIP) establishes (Article 19) that "States shall consult and cooperate in good faith with the indigenous peoples concerned through their own representative institutions in order to obtain their free, prior and informed consent before adopting and implementing legislative or administrative measures that may affect them."
Rights to control and management of lands and resources	Through customary institutions and laws.
Rights to development and equitable benefit-sharing	Including to determine the development or use priorities and strategies on their lands, territories and resources and to benefit equitably from conservation and sustainable use of such areas and resources.
Rights to traditional knowledge and indigenous heritage	
Redress	UNDRIP includes provision for redress for deprivation of indigenous peoples' means of subsistence and development, and for lands taken without free, prior, informed consent.

2.4 Conservation Initiative on Human Rights from World Parks Congress in Durban

In 2003, when the protected areas community and a wide variety of stakeholders convened in Durban, South Africa for the V IUCN World Parks Congress, they came prepared to address critical challenges confronting protected areas at that time. Protected area administrators and agencies, NGOs, governments, funders, scientists, and – most significantly – indigenous peoples and community leaders from around the world gathered at the Congress to debate the state of protected areas and the difficulties they face, as well as the path forward. Three of the numerous results of the 2003 Durban Congress are particularly important. First, the Congress provided a once-in-a-lifetime chance for a wide-ranging and candid dialogue between indigenous peoples, local communities, and conservation agencies and organisations. Community members discussed how protected places have impacted their lives and advocated for a more people-centered conservation strategy. By the end of the Congress, there was widespread agreement that indigenous peoples and local communities must be included in future decision-making processes based on respect for their rights in order to promote mutually reinforcing park policies and development methods. The conservation community also pledged to improve their interaction with local stakeholders, including examining policies and procedures related to preserving indigenous peoples' and communities' rights and interests. Second, there's the agreement, which is a non-binding pledge aimed at guiding activities over the following ten years. While the Accord praised local conservation efforts, it also noted that indigenous peoples and local communities – as well as their contributions to conservation – needed recognition and protection in general. The Accord, as well as the Action Plan that accompanied it, asked for commitment and specific actions to rectify such oversights. Finally, Indigenous and Community Conserved Areas (ICCAs) were established based on a Durban WPC Recommendation, and they were later accepted by the Convention on Biological Diversity. ICCAs are 'natural and/or modified ecosystems containing important biodiversity assets and ecological services, voluntarily maintained by (sedentary and mobile) indigenous and local populations, through customary laws and other effective means', according to the recommendation. Community participation in the Congress, recognition that the traditional protected areas model had flaws in consistently addressing indigenous peoples' and local communities'rights based on international standards, and conservation organisations' desire to improve practise all contributed to defining and clarifying key elements of the new conservation approach that emerged from the discussion.

Since the 2003 World Parks Congress, the movement to address human rights problems in the conservation context has grown in intensity (WPC). In the years after, at least 20 IUCN resolutions and policy decisions on conservation rights have been enacted, for example. The United Nations General Assembly approved the Declaration on the Rights of Indigenous Peoples (UNDRIP) in 2007 after 20 years of discussions. The statement, while not legally enforceable, reflects globally accepted legal principles and serves as the minimal bar for safeguarding indigenous peoples' individual and communal rights. Many of the rights stated in the document, such as self-determination, land rights, free, prior, and informed consent (FPIC), and the ability to choose one's own development path, have significant implications for natural resource management. UNDRIP became an essential benchmark for conservation groups to build their institutional rules on, as well as a tool to guide action. Its acceptance was a great success and a worldwide monument to indigenous peoples' human rights for indigenous peoples all across the world. These regulatory advancements have provided the framework in which conservation groups have taken significant steps to include human rights respect into their collective practise. The Conservation Initiative on Human Rights was formed two years after UNDRIP was adopted by the eight largest international conservation organisations (CIHR). This organisation is dedicated to improving their own human rights standards as well as creating and advocating best practises for the whole conservation community. The organisation works together to promote the beneficial links between conservation and people's rights to secure their livelihoods, enjoy healthy and productive surroundings, and lives in dignity, all motivated by a shared goal. They are dedicated to four fundamental principles that govern the integration of human rights into the policies and activities of each business. These guidelines are outlined below (Table 2.2). The CIHR member organisations have made concrete efforts to integrate and protect human rights to develop overarching policy for enhancing practise on the ground, all based on these principles(CIHR, 2014) (Table 2.3).

Biodiversity conservation can be treated as a commons problem, specifically as a multilevel commons problem. Biodiversity is a global common important for humanity as a whole, a regional common important for ecotourism and other benefits, and a local common that produces ecosystem services for human well-being at the community level. As a multilevel common, the ownership and control of

Table 2.2 Principles of CIHR

	Principles	Elucidations
1	Respect human rights	Respect internationally proclaimed human rights; and make sure that they do not contribute to infringements of human rights while pursuing their mission.
2	Promote human rights within conservation programs	Support and promote the protection and realization of human rights within the scope of their conservation programs.
3	Protect the vulnerable	Make special efforts to avoid harm to those who are vulnerable to infringements of their rights and to support the protection and fulfilment of their rights within the scope of the conservation programs.
4	Encourage good governance	Support the improvement of governance systems that can secure the rights of indigenous peoples and local communities in the context of their work on conservation and sustainable natural resource use, including elements such as legal, policy and institutional frameworks, and procedures for equitable participation and accountability.

Table 2.3 CIHR member organizations and relevant progress

CIHR member organizations	Progress
BirdLife	The BirdLife partnership of 121 national NGOs approved a revised vision, which confirmed BirdLife's commitment to equity, empowerment and sustainable livelihoods. The BirdLife Partnership's vision is of "a world where nature and people live in greater harmony, more equitably and sustainably," and its commitments include "to sustain the vital ecological systems that underpin human livelihoods, and enrich the quality of people's lives" and to "empower people and contribute to the alleviation of poverty."
Conservation International (CI)	Adopted a new mission that better links the conservation of biodiversity to human well-being.
Fauna & Flora International (FFI)	FFI embarked on a four-year, donor-funded program to strengthen its own and partners' capacity to "deliver effective livelihood interventions which reduce both poverty and threats to biodiversity across its project portfolio." A Biodiversity and Human Needs (BHN) team was established to facilitate project teams' access to expertise and opportunities for capacity building and shared learning.
International Union for the Conservation of Nature (IUCN)	IUCN has been actively contributing to ensure greater respect for the rights of indigenous peoples and local communities in conservation and specifically in protected areas, both in policy frameworks and on the ground.
The Nature Conservancy (TNC)	The nature conservancy adopted a new conservation framework: "Global Challenges, Global Solutions," to better demonstrate how conservation is essential to improving people's lives. The nature conservancy is currently updating its "conservation by design" methodology to emphasize the interconnectedness of people and nature, including a systematic approach to integrating human well-being into project design and monitoring and evaluation, with a focus on people's empowerment.
Wildlife Conservation Society (WCS)	WCS has generally had positive relationships with local people affected by our work in our long-term conservation landscapes, and some of the field programs have been leaders in introducing a rights-based approach to doing conservation.
World Wide Fund for Nature (WWF)	As a founding member of the CIHR, WWF has taken a number of steps to build institutional knowledge, internal policy, and external engagement to ensure that we proactively build human rights and social development into our conservation work A set of social policies guide the integration of social dimensions in WWF's conservation work: the Statement of Principles on Indigenous Peoples and Conservation (agreed 1996, updated 2008), the Policy on Poverty and Conservation (2009), the Conservation Initiative on Human Rights Framework (2009), and the Gender Policy (2011). Building on diverse initiatives around poverty, social equity, and indigenous peoples' rights, the Social Development for Conservation (SD4C) team was established in 2009 to build coherence, effectiveness, and support for capacity across WWF's Network.

biodiversity are complex. Some biodiversity is under state ownership, some is under the control of communities, and some is privately owned. Many of the lands that support biodiversity are under multiple and competing claims, including nominally government-owned forests under community control; protected areas designed for biodiversity conservation but that allow for human use, as with the World Conservation Union's protected-area categories V and VI and locally maintained traditional protected areas, such as the sacred groves of Kerala, India, which can be as effective as nearby government protected areas. If the ecosystems that support biodiversity were simple systems, and if the creation and implementation of protected areas did not involve social and political controversy, state control would be an appropriate low-cost solution. But this is not the case. Conservation is typically a complex systems problem, because the natural environment itself is a complex adaptive system with issues of scale, uncertainty, and multiple stability domains. Ecological systems are hierarchically organized, with each subsystem nested in a larger subsystem. Complex systems theory holds that the levels are linked, but that each level requires diverse concepts and principles. Self-organization provides a unifying principle for complex adaptive systems: "The specifics are in the often simple-rules that govern how the system changes in response to past and present conditions, rather than in some goal-seeking behaviour". The social systems involved in conservation also are multilevel, with institutions at various levels of organization from local to international. Processes at these levels require different but overlapping sets of concepts and principles, an idea reflected in the common's literature. Because each level of a scale is different, the perspective from each level is likely also different. The global lens of biodiversity conservation (that it is a global commons) is therefore different from the local lens on biodiversity (local commons for livelihoods). This difference does not mean one perspective is right and the other wrong; it means they can both be correct from different points of view. Pluralism in perspectives is mirrored in pluralism in knowledge. In conservation disputes, local knowledge may often appear at odds with science. But in many cases, the differences in knowledge and understanding of a resource system have to do with differences in the level at which information is obtained, a point often missed in blueprint approaches. Insights from the path-dependency concept in complex systems further reveal the inadequacy of the blueprint approach. As may be applied to social sciences, this is the argument that context (history, politics, and culture) is important in understanding a particular case. This is not to say that changing direction is impossible; rules and practices change all the time through adaptation and learning, although in some cases changing directions or reversing a path may be costly. Rather, the point is that each case is conditioned by the context in which it developed, meaning a solution package developed from one case cannot readily be transferred to others. For example, the community-based marine protected-area approach, developed in one area of the Philippines and replicated throughout the country with little attention to context, resulted in a high rate of failure. However, with attention to context, it should be possible to transfer lessons, in so far as similar diagnoses call for similar treatments. In sum, the perspective of biodiversity conservation and the relevant social–ecological system as complex and multilevel is in sharp contrast with the simple view of biodiversity conservation that has led to blueprint solutions (Berkes, 2007).

2.5 Gender Equality

Eliminating discrimination against women the rights of women to enjoy human rights and fundamental freedoms on a basis of equality with men have also received attention in international law, and are relevant to conservation contexts. Women often have distinct roles and needs in relation to natural resource management, but also suffer from discrimination that can limit participation in decision-making and control over natural assets (such as land). The Convention on the Elimination of All Forms of Discrimination against Women (CEDAW, adopted in 1979) is the main international instrument addressing the equal rights of men and women. It includes provisions specifically oriented towards rural women to "ensure, on a basis of equality of men and women, that they participate and benefit from rural development," including through rights to: participate in development planning, obtain equal access to economic opportunities and development programs, and receive equal treatment in land and agrarian reform as well as land resettlement schemes.

2.6 Key Issues at the Intersection of Conservation and Human Rights

While an innumerable of human rights related issues can arise in conservation, there are some particularly common and/or challenging issues that call for attention are (Table 2.4):

2.7 Frameworks for Discussion of the Crucial Topics

There are several contexts in which rights issues are most prominent or likely to emerge. Rights issues are relevant for conservation work at multiple levels. Often rights linkages are most clear on on-ground work, but are also relevant to activities where they may be less visible, such as broader spatial planning, addressing drivers, influencing policy and governance frameworks, and establishing financing mechanisms for conservation activities. Integrating rights issues into these broader frameworks also helps to promote enabling conditions for on-ground work, and 'scaling up' of rights-based approaches to conservation (Table 2.5).

Scoping key issues at the intersection of conservation and human rights, and guiding principles associated with them help to develop relevant policies and/ or guidelines on human rights issues in conservation, particularly among organizations participating in the Conservation Initiative on Human Rights. At the same time, promotion of dialogue and action among conservation organization staff, rights-holders, experts and other partners on experience and "best practices" for integrating human rights in conservation in particular contexts. Issues that could be taken

Table 2.4 Key issues at the intersection of conservation and human rights

Key Issues	Elucidations
(i) Participation in decision-making	Participation is a procedural right that is also widely recognized – e.g., in principle 10 of the Rio declaration – To contribute to "better" (more informed, fair, and equitable) decisions. Participation is also an important component of good governance, itself increasingly recognized a prerequisite for effective and equitable natural resource management.
(ii) Free, prior, informed consent	Free, prior and informed consent (FPIC) refers to the principle that communities have a right to give or withhold consent to actions that will affect them, particularly actions affecting their customary lands, territories and natural resources One of the most prominent issues at the intersection of conservation and human rights is weak tenure, and especially the gap between statutory and customary rights to lands and resources. In many parts of the world, governments (often colonial governments) claimed as state land areas that were traditionally owned and governed by indigenous peoples and local communities, often as common property systems.
(iii) Tenure security, especially conflicts between customary and statutory tenure	One of the most prominent issues at the intersection of conservation and human rights is weak tenure, and especially the gap between statutory and customary rights to lands and resources. In many parts of the world, governments (often colonial governments) claimed as state land areas that were traditionally owned and governed by indigenous peoples and local communities, often as common property systems. Where people have maintained continuous attachments to their lands, there are likely to be situations of "legal pluralism" – i.e., overlapping systems of statutory tenure (codified in state law) and customary tenure (derived from ancestral ties and regulated through traditional institutions). Community-based management of forests, wildlife, and other natural resources offer alternatives to state management activities that have often been ineffective in stopping deforestation and degradation. Clarity of tenure decreases contestation and conflict over lands and resources. Conversely, lack of understanding or attention to how conservation activities may impact tenure increases risks of territorial and land rights being negatively affected, for example in relation to state-managed protected areas.
(iv) Cultural rights and bio-cultural diversity	There are often strong positive synergies between rights of communities to their cultures and biodiversity conservation objectives. Perpetuation of many traditional cultures depends on maintenance of healthy natural systems and species diversity. Traditional governance institutions and practices, knowledge, beliefs and values also often contribute significantly to maintaining and enhancing natural systems. These inter-linkages between peoples' ways of life and their environments, and the role of cultural groups in valuing, stewarding and enhancing biodiversity are increasingly referred to as biocultural diversity. As one example of these broader cultural values and links, many cultures maintain sacred areas such as forests or mountains that have historically protected these areas, even if for very different reasons and with different understandings than those of outside observers.

(continued)

Table 2.4 (continued)

Key Issues	Elucidations
(v) Sustainable development and equitable benefit-sharing	For many rural people, a healthy environment is a primary foundation for development89 and generation of economic benefits. Thus, there are strong motivations for rural people to use natural resources in ways that maintain these key "assets" for the future. One way that sustainable development and conservation may be made more compatible is through the generation and equitable sharing of *increased* benefits from natural resources, through 'green economy' and alternative economic growth paths, or (often in more limited ways) from additional income streams linked to protected areas or payments for environmental services.
(vi) Displacement and restrictions on resource access	Displacement has been one of the most contentious issues at the intersection of conservation and human rights. Displacement is commonly understood as physical relocation, but can also be economic – As when access to resources that people depend on for livelihoods is restricted, affecting enjoyment of basic rights (e.g., food, health). The right to adequate housing and the right to "choose residence" have also both been interpreted to protect against arbitrary or forced/unlawful eviction, though the specific implications of these interpretations for conservation require further exploration.
(vii) Law enforcement	Enforcement of laws and regulations governing biodiversity protection and sustainable use of natural resources are one necessary element of conservation and resource management. Law enforcement protects natural systems, the species that inhabit them and the rights of people who own use and/or depend upon them. Good governance requires application of the rule of law, including its fair and just enforcement. At the same time, some of the sharpest conflicts between conservation and human rights concerns are associated with law enforcement activities.

up as a focus of more in-depth learning and exchange through CIHR activities, and provides a foundation for future work of the Initiative in support of measures by participating organizations to enhance capacity, develop program integration tools and establish accountability mechanisms. As this work moves forward, some lessons from international development organizations that have implemented rights-based approaches include that such integration takes substantial time, financial resources, and technical capacities, including expertise and understanding that organizational staff may not readily possess; and respecting rights in some cases will mean difficult trade-offs with conservation objectives- though positive synergies are also often possible. These challenges indicate that comprehensive and effective rights integration will require new resources, strategic partnerships, and deep organizational commitments that will take time to take root. Organizations must develop their own paths towards rights integration, including determining how to make the 'bottom up' and 'centre out' meet in effectively in addressing complex, localized rights issues. Documenting adherence to rights principles will be essential, both for internal efforts, and to support commitments to transparency and accountability in relationships with other actors including communities, governments, donors and the interested public. Finally, ongoing learning from experience will be needed to improve and adapt approaches to addressing human rights issues in diverse settings over time (Springer et al., 2011).

Table 2.5 Context to appear the key issues

Context	Key issues
Protected areas	Historically, protected areas are the conservation context in which human rights concerns have been most prominent. In recent years, this conflict has been addressed to some extent by greater attention to improving the quality of protected areas governance, including respect for human rights, transparency, accountability, participation, equity and benefit sharing. There has also been broader recognition of and support for diverse governance arrangements. Today, it is commonly recognized that protected areas may be governed by governments, by private individuals or companies, by communities (e.g., indigenous and community conserved areas – ICCAs), or by combinations of these (i.e., co-management). In relation to tenure/ resource rights, participation and FPIC, cultural rights and bio-cultural diversity, displacement and restrictions on resource access, sustainable development and equitable benefit-sharing and law enforcement.
Climate mitigation (REDD+)	Discussion of human rights – Including rights of indigenous peoples – Has been prominent in development of frameworks for REDD+ because of concerns that REDD+ implementation could adopt exclusive, centralized management approaches (similar to those characteristic of 'traditional' protected area management), or otherwise adversely impact vulnerable forest communities. Participation in decision-making about REDD+ reflected recognition that full and effective participation of all stakeholders is necessary to develop REDD+ programs that will minimize adverse impacts and generate broad climate, biodiversity and social benefits. FPIC enables communities to guard against negative impacts, and shape REDD+ initiatives to increase their positive benefits. Land and resource rights will also substantially influence who will receive benefits from REDD+. The role and contribution of indigenous peoples' traditional knowledge, innovations and practices has been emphasized in relation to development of international and national frameworks for REDD+. A key source of interest in REDD+ is its potential to generate benefits for indigenous peoples and local communities, as historic stewards of forests. However, benefits will depend on financial mechanisms that prioritize incentives for forest communities, and effective governance of those mechanisms to ensure that benefits are equitably distributed. REDD+ safeguards, such as for full and effective participation, FPIC and protection of customary tenure, are largely aimed to prevent displacement or involuntary restrictions on resource use that might otherwise result from schemes to reduce emissions by curtailing forest use.

(continued)

Table 2.5 (continued)

Context	Key issues
CBNRM	In general, Community-based Natural Resource Management (CBNRM) has emphasized positive approaches that link community rights and livelihoods with enhanced conservation, including by building on traditional management systems and knowledge. At the same time, changes in resource use and new formal obligations associated with CBNRM policies or regulations mean that issues may arise in relation to participation in decision-making- particularly ensuring that community participation is inclusive and does not result in negative impacts on more vulnerable members and/ or 'elite capture' of benefits. FPIC is particularly important where a CBNRM initiative, such as community forestry, involves changes in tenure (e.g., registering an area under a government community forestry lease), new land and resource management obligations, and/or changes in access to and use of resources. CBNRM often contributes to greater security of land tenure and resource rights. CBNRM may support cultural values and practices but may also approach resource management in particular ways that do not accord with cultural perspectives. Main objective of CBNRM is to contribute to community development by generating benefits from sound management of natural resources, initiatives also often imply additional costs in terms of management and forgone use which need to be balanced with benefits.
Species conservation	Other than main strategy for species conservation in protected areas there are other human rights issues that arise in the context of species conservation activities taking place outside protected areas. Participation issues may arise in relation to policy decisions about species protection measures, especially those that may be significant for subsistence or cultural heritage, as well is in decision-making about species protections in particular places. Many traditional peoples have strong cultural ties to specific wildlife species as well as traditional knowledge, which may support or be impeded by conservation measures. Devolution of rights to wildlife to communities has contributed to sustainable development under community-based wildlife management programs; benefit-sharing issues also arise in the context of wildlife management on traditional lands and human-wildlife conflicts. Species protection measures such as anti-poaching activities (within and outside protected areas) are a primary context for conflicts associated with law enforcement.

(continued)

Table 2.5 (continued)

Context	Key issues
Private sector engagement/ commodity certification	Human rights issues surrounding conservation engagement with private sector actors are often particularly fraught, due to the negative human rights record of many industries – Especially extractive industries– That have operated in high cultural and bio-diversity areas, and the resulting mistrust with which many indigenous peoples, local communities and human rights organizations view them. Historically, participation of indigenous and local communities in decision-making about industrial commodity production in their areas has been very weak. Free, prior, informed consent developed primarily as a response to the devastating impacts on communities that have occurred as a result, and is gradually becoming adopted as a best practice standard in industry sectors. Industry concessions (oil/gas, mining, logging) are often granted in areas that are formally considered to be state lands but overlap with customary claims– Resulting in conflicts and weakening customary tenure. Industrial operations can contribute to local economic development and share benefits; however, development may not be sustainable or benefit-sharing equitable. Large-scale infrastructure and industrial developments have been significant drivers of displacement over time, leading to establishment of involuntary resettlement standards that aim to minimize negative impacts. Issues that arise in relation to industry include both the imposition of restrictions on resource access and use contrary to customary rights, and the use of force to guard and secure local operations. The implications of these issues for private sector engagement by conservation organizations are twofold. On the one hand, engagement– Such as to develop and support implementation of commodity certification and other good practice standards– Offers opportunities to further human rights, through adoption and implementation of rights-based standards on FPIC, tenure, equitable benefit-sharing and other key issues.

2.8 Community Led Conservation

In India, forests support, directly, more than 100 million forest dwellers living in and around forestlands and indirectly, another 275 million. The genesis and evolution of forest policies in India shows that the Forest Policy of 1894 was laid with an objective to manage the forest to promote the general wellbeing of the country, preservation of climatic and physical conditions, fulfilling the needs of the local population and realisation of maximum revenue from the forests without any role of the community. Subsequently, the National Forest Policy of 1952 was formulated which emphasis covering of one-third of India's land area under forest cover for the ecological wellbeing of the country, maintenance of forests to address the needs of defence, communication and industries. The local communities were alienated in management, which led to commercial exploitation, incidents of encroachment and degradation of forests. The social forestry programme in the 1980s and the recognition of Community Forest Management in several parts of India, was an important catalyst to formulate the National Forest Policy of 1988, which was radically different from previous policies. Forests were meant to conserve soil and the environment and meet the subsistence requirements of local people, and recognised community role in the management of forests. To facilitate implementation of the 1988 policy, the 1990 joint forest management (JFM) Guidelines were passed, which envisaged the management of forests jointly by the local community and the Forest Department, by formation of Joint Forest Management Committees (JFMCs). Since then, 28 states have adopted the JFM resolution and the Ministry of Environment and Forests has subsequently issued 2000 and 2002 guidelines to strengthen sustainable forest management and equitable community participation in JFM. In India, 17.33 million ha of forests are being managed through 84,642 JFMCs in 28 states (MoEF, 2003).

2.9 Community Forestry: Case Studies from South and Southeast Asia

The impact of community forestry is difficult to determine on a regional basis due to lack of information on national level monitoring and evaluation. In Southeast Asia, much of the experience emerging in rural areas is captured only in occasional case studies. It is possible, however, to examine a variety of examples of community forestry systems in different environments to assess their impact on forest cover and biodiversity.

2.9.1 Indonesia

A study of Damar (*Shoreajavanica*) forest gardens, owned and managed by Krui families, found that the damar forests had better biodiversity compared to rubber estates. There were 230 plant species in adjacent rain forests, 120 in damar forests, and only ten species in the rubber estates. Further, it is an important habitat for endangered mammals such as the Sumatran rhinoceros, the Sumatran goat, tigers, tapir, gibbons and siamangs (monkeys). In effect, the damar forests act as a critically important buffer zone to the Bukit Barisan Selatan National Park, greatly extending the habitat for many species. Maintenance of Damar forests have conserved more than 50% species in rain forests, and in addition provided over three months of employment with substantial income (Poffenberger, 2006).

2.9.2 Vietnam

In Cao Bang Province to the north, the Nung, an ethnic community in limestone forests, planting a combination of indigenous pioneering valuable timber species with natural regeneration, have reforested limestone hillocks. The restoration of the limestone forests has facilitated the reestablishment of spring flows that provide water for the lowland rice fields. It has also allowed for the return of many indigenous mammal species, including five endemic and 26 rare species. Thus, in addition to restoration of degraded hillocks, the community efforts have brought back endemic and rare species in their forests.

2.9.3 Cambodia

In KompongPhluk village, community members have been protecting flood forest for nearly 60 years. At present, the village controls over 15,906 hectares of land formally recognised by the provincial government. The community follows a resource management plan, allowing controlled fuelwood harvesting, monitoring fishing gear and catch levels, and generating fees for management activities. With over 200 different species of fish in the lake, many endemics, the flood forests protected by the communities provide a critical habitat for biodiversity conservation. The community effort is helping in expansion of both mangrove and freshwater forests that are critical aquatic ecosystems.

2.9.4 Philippines

Community based forest management (CBFM) in the Philippines includes management of tropical forests through enrichment planting, timber stand improvement, limited harvesting, rehabilitation of degraded lands through reforestation, assisted natural regeneration, and agroforestry. About 6 Mha of forest lands are under CBFM of which 4.7 Mha have been issued with various forms of land tenure instruments including 1.57 Mha with Community Based Forest Management Agreement (Lasco & Pulhin, 2006). About 6,90,687 households are involved in the implementation of the CBFM programme with around 4.14 million people who are potential direct beneficiaries of the programme. In the CBFM site of Cebu (Central Visayas), conservation of valuable habitats of the endemic life found in the area has promoted biodiversity, which include 122 endemic species of birds, 27 species of mammals and 27 species of reptiles and amphibians. On the negative side, tree plantations have been blamed for clearing of natural forests to give way to plantations, resulting in loss of biodiversity. The biomass productivity rate varies from 20 t/ha/year in good sites to 1 t/ha/year in degraded areas.

2.9.5 India

In India, large extents of degraded forest area brought under the JFM programme has led to improvement in forest cover (FSI, 2003), tree density, biomass and biodiversity(Murali et al., 2002; Ravindranath et al., 2000). Thus, the aim of JFM, that participation of local communities to enhance forest cover, seems to be bearing fruit, surely though slowly. Here we discuss the ecological improvements, as observed in different provinces namely: Andhra Pradesh, Gujarat, Karnataka, Rajasthan, Tripura and West Bengal and in India. JFM in India originated from Arabari range of Midnapore division in West Bengal. The experiment during 1974 was started primarily to ward off the distress condition of local people and to join hands with the Forest Department to improve degraded areas where people were given a stake in the profits resulting from protection. The experiment began to change the overall policy of state-controlled forests to joint management in India. In Rajasthan, Gujarat, Karnataka and Tripura provinces, indigenous community forestry systems existed even before national policy initiatives. Thus, the policy initiatives came in handy to rejuvenate the already existing traditions (Ravindranath et al., 2006).

2.10 Social Forestry

Southeast Asia is currently pushing forward a revived social forestry (The term social forestry as this is commonly used in policy documents through much of Southeast Asia in this context social forestry is often used interchangeably with community forestry, community-based forest management and similar terms.) agenda. Indonesia has set a target of 12.7 million hectares (m ha) of forest, approximately 10% of the state forest estate, to be managed under various formal social forestry schemes by 2019, and official statistics suggest that the Indonesian Ministry of Environment and Forestry (MOEF, 2018) has allocated permits for 3.4 m ha as of September 2019. Vietnam set a target of 4 mil ha for social forestry which it met in 2016, four years ahead of its target date (RECOFTC, 2018) and of which, most are being managed by commune people's committees (MARD, 2017). In Malaysia, Sabah's Forest Policy 2018 emphasizes sustainable forest management (SFM) principles to generate diverse revenue streams for the state and benefits local communities (Sabah Forestry Department, 2018), with social wellbeing objectives that aim 'to strengthen the participation local communities in forest management activities' and 'to encourage Native [people's] participation in forest industries'.

2.10.1 Social Forestry and Agrarian Reform in Indonesia

The presidency of Joko Widodo, which began in 2014, introduced margins. These included agrarian reform programs, of which Social Forestry (SF) is a major component. This reinvention of SF promoted the idea that by providing access to land, business opportunities and vocational training, poor rural farmers and forest communities will be able to escape poverty (MOEF, 2018). Widodo's target of 12.7 million ha of forest area under SF by 2019 has led to an explosion of SF permits. While there are varying figures, official statistics suggest that 1.27 m ha were allocated to various SF schemes between 2015 to June 2018 (MOEF, 2018), with a large jump to 3.4 m ha by Sept 20,195. This newest iteration of the SF policy is distinct. In the 1970s and 1980s, SF was formulated as a 'development' program for communities living in or around forest plantation concessions to resolve land conflicts and social discontent to further the interests of state and private corporations. This changed in the 1990s and 2000s with a series of ministerial regulations and national forest laws that laid out formal SF schemes for the devolution of rights and access aiming to empower communities to manage forests (Moeliono et al., 2017; MOEF, 2018). The current definition of SF is: 'a system of sustainable forest management within the state forest area or within privately owned or traditionally (adat) owned forest, whereby local communities or adat communities are the principal actors carrying out forest management to improve their livelihood, maintain environmental balance and their social-cultural heritage in the form of Village Forests, Community Plantation Forest, Community Forest, Traditional [or Customary] /Adat Forest and Partnership Forests". Officially, the goal of SF remains to solve tenurial conflict and achieve equity for local and adat communities living in or around forests for their wellbeing and sustainability of forests' (Ibid, article 2.2). SF is implemented according to the five formal schemes as per the Ministerial Decree P.83. Three schemes are based on management permits issued to village enterprises (Village Forests), communities/groups (Community Forest) and individual or cooperative plantations (Plantation Forest), one scheme regulates co-management of forests in protected areas (Partnership Forest), and only one recognizes full ownership of forests by customary or 'adat' communities (Adat Forest). However, the process to obtain the certificate of adat ownership involves the laborious process of gaining official recognition as 'adat' community by the district government and a territorialization process that can exclude certain groups and their claims (Myers et al., 2017). Further, adat forests comes with the full responsibility to protect the forest but with little government support to do so (Fisher et al., 2018). Invisibility of local forest governance practices. Although official SF policies responsibilize local communities to manage forests, participation at the community level is often weak or limited to village elites (De Royer et al., 2018). Further, the SF territory comes with state-imposed restrictions on sites, use and access (De Royer et al., 2018; Erbaugh, 2019), thus making impossible for diverse local governance practices and informal rights long practiced in many forested sites to be recognized by, or drawn into, formal social forestry structures (Bong et al., 2019). This incompatibility has created conflicts and contestations over use and access (Cummins & Yamaji, 2019). Despite the restrictions, local communities are still inclined to obtain SF permits for achieve some security of access to the forest land and resources they are dependent on (Myers et al., 2017). Bureaucracies, politics and interests. To obtain a SF permit, communities have to develop and submit a technical proposal documenting required criteria such as proof of a community institution or cooperative to manage the forest, the boundaries of the area to be managed, and an approved management plan (Moeliono et al., 2017). This territorialization process excludes local groups who may not fit, or choose not to participate, in these 'neat' community institutions or cooperatives, or those who may demand more from or contest the specific discourses of forest stewardship within the formal SF policy (Myers et al., 2017; Sahide et al., 2020). There is also discontent amongst local (district) agencies who used to play substantive roles in forest governance and now excluded under the new Law 14 on Regional Government (Fisher et al., 2018). In addition to SF, Indonesia's push for agrarian reform is spearheaded by the policy on Identification of Land for Agrarian Reform policy (Identifikasi Tanah Obyek Reforma Agraria, TORA). TORA involves the certification of 4.5 m ha of land informally managed by individual farmers, the re-distribution of 4.1 m ha of state forest lands, and the provision of use rights for 0.4 m ha of idle or abandoned lands (Resosudarmo et al., 2019; KSP, 2016). The latter two categories are forest areas that are earmarked for crop plantations, unproductive forests slated for conversion and potential areas for paddy rice farming (KSP, 2016). This has implications for potential disputes between the MOEF and the Ministry of Agrarian and Spatial Planning (Sahide & Giessen, 2015). SF permits are being awarded or planned for 12.7 mil ha of the State Forest Area while these same areas are also being 'released' as private property to communities and individuals for farming. While contestations over TORA have not provided much space for the involvement of civil society organizations (CSOs), the SF policy process on the other hand has embraced their participation. A loose multi-stakeholder platform, established since the 1990s of CSOs, academics, individuals and businesses actively engaged in all aspects of social forestry, was institutionalized as the formal 'SF Acceleration Working Groups' (POKJA PPS) through a decree by the Director General of Social Forestry (P.14/PSKL/SET/PSL.0/11/2016) to support fast-track implementation of SF schemes on the ground. This act effectively subdued the platform's critical dissent of the policy by rendering its members as technical experts and agents to implement the SF policies. The POKJA PPS engages in the entire territorialization process of SF at the local and provincial levels, including facilitating permit applications, supporting the administrative and biophysical verification and monitoring processes, and reports to the provincial governors (MOEF, 2018; Resosudarmo et al., 2019) again bypassing local government agencies. In interviews, members of the multi-stakeholder platform suggest that priority is to first secure SF permits, and then work on other equity issues later. In addition, SF Community Business Groups were formally established to assist in the development of social forestry enterprises (MOEF, 2018), which are considered as a successful outcome of a SF scheme. The Indonesian assemblage with its diverse actors, interests and high levels of interconnectedness and contestations suggest that different possibilities could be negotiated at specific moments and places.

2.10.2 A State-Driven Social Forestry in Vietnam

The Vietnamese Constitution defines the State as manager of all forests and land. Insights from interviews with government actors between 2016 and 2018 suggest that social forestry in Vietnam can be understood in two different ways: social forestry and community forestry. The term 'community forestry' implies attention to communities as the focus of the policies while 'social forestry' focuses on mobilizing finance from private sectors and social actor groups. Findings from interviews conducted with central and provincial government actors as well as from reviews of policy documents suggest that the central government emphasizes 'social forestry' and forest policies prioritise collaborative models with local communities and households through an economic and production lens. Moreover, local governments tend to apply a 'community forestry' approach in their forest protection and development plans emphasizing support to and empowerment of farmers, households, and communities to manage forests. The distinction between the two terms suggests differentiated priorities for local actors in the management of the state's forest estate. The forestry sector, as with other sectors, is managed by the State to align with the political interest of the Communist party, whose views towards social forestry are not static. Following independence from the French in 1945, forests were exclusively managed by the State through its state forest enterprises (Directive No. 15 CT/CTCW 1961). The early 1970s saw emergence of a political shift towards collaborative forest management (Forest Protection Act 1972), which led to the coercion of local people to support the state as temporary laborers for state forest enterprises often with no direct benefits in return (Pham et al., 2012). Decision No. 184-HĐBT 1982 and Forestry Development Strategy 1986–2009 recognized the role of non-state actors (cooperatives, communities and individual households) to manage forests but without accompanying rights. In the late 1980s, transformation of the country's economy from state central planning to market orientation, DoiMoi, triggered liberalization in the agriculture sector and large-scale changes in household ownership of products and land, but this was not translated to the forest sector with "ownership" still limited to use rights. The processes of territorialization with selective inclusions and exclusions is to facilitate economic production. Significantly, the open-door policy of DoiMoi enabled foreign organizations to introduce new narratives and demands for devolution and participation (Dang et al., 2012). Increasing openness but not for all. Early manifestations of social forestry started in the 1990s (Forest Protection and Development Law 1991 and Land Law 1993) emphasized local communities' duties rather than the benefits they could gain from managing forests (Clement & Amezaga, 2008). Many central and provincial government interviewees claimed that 'In the past, no one wants to protect and manage forests but central and provincial government assign forests to them as political tasks. Local people saw this as a political mission rather than for their own benefits'. In forest areas belonging to state forest enterprises, local people were contracted to work on forest protection or planting (Decree 01/1995/ ND-CP, Decision 661/1998/ QD-TTg), with a share of benefits depending on time and labor invested (Ngo & Mahdi., 2017). Community forestry was officially recognized in the Forest Protection and Development Law 2004 where 'ownership' rights of non-state actors were specified, and the first state-led community forestry pilots were implemented in 17,000 ha across 64 villages in 2007 (RECOFTC, 2014). These pilots drew on lessons from the legacy of past forestry policies and projects and the contestations over changing rights and exclusions (Sikor & Nguyen, 2011). The government considered community forestry to be a solution to the conflicts between the state and local people in forest and land governance, between conservation and development practices, and an opportunity to address high rate of deforestation and degradation (Decree No. 02/1994/ND-CP, Decree No. 01/1995/ND-CP). The expansion of collaborative forestry approaches in Vietnam was in large part shaped by the policy narratives of international actors, through the influence of bilateral aid programs on social forestry funded by Switzerland (1992–2002) and the Trust Fund for Forest funded by Finland, Netherlands, Sweden and Switzerland (2004–2015). In particular, MARD interviewees (2018) attributed the government's shift towards social forestry evident in the Forest Protection and Development Law 2004 to the Swiss-funded program on social forestry. Declining state budget allocations to the forest sector pushed the government to consider alternative financing from non-state actors (Pham et al., 2008; Pham et al., 2013), leading to the development and implementation of the national Payment for Forest Environmental Services policy (PFES Decision 380 and Decree 99). International donor-funded programs on PES (Germany and the US) and REDD+ programs (Norway) set the tone on the need to engage meaningfully with communities, indigenous people and private sector. However, State dominance in these processes leaves limited political space for non-state actors, e.g., NGOs and CSOs, in Vietnam to exert an influence on the final policy outputs (Pham et al., 2014). The gap between 'community forestry' and 'social forestry'. The Forestry Law 2017 (enacted on 1/1/2019) gave formal recognition of religious and customary rights over forests for the first time in Vietnam's legal framework. While policies have increasingly emphasized the rights and important roles of communities, individual households and private sector as forest owners and managers, this shift remains rooted in the government's interest to ensure sustainable revenues for the forestry sector – thus reflecting the disparity between the Vietnamese meanings of 'social forestry' and 'community forestry'. For example, social forestry as defined in the PFES policy is a process where all actor groups contribute their financial resources to protect forests and reduce the burden on the state budget (Pham et al., 2013). The Vietnam Forestry Development Strategy (2006–2020) emphasizes an economic and entrepreneurial approach in all forestry actions with local communities and households to advance state development objectives. However, forest "ownership" still only refers to limited forest use rights (except in plantation forests where households have tenure) and little autonomy in decision-making related to their forests (Dang et al., 2012). An international donor agreed (2019): "Yes, you might see community forestry and social forestry everywhere in policy documents and in the speeches of policy makers, the fact that the state and its state forest enterprises, protected areas and national park and state management boards own and manage more than 85% of forests, mostly the best quality forest in Vietnam while communities and households receive the most degraded forests pose significant challenges for non-state actors to engage in forestry protection and development". By the end of 2018, less than 2% of total forest had been officially allocated to communities (Vietforest, 2018; Wong et al., 2020). While there are claims and forms of resistance at different levels, the participation of local people and CSOs in formal SF planning and implementation processes remains limited, despite the participatory rhetoric in policies (from interviews with CSOs in 2019) thus highlighting the state's dominance in the Vietnam assemblage, which effectively dictates territorialization processes to maximize its benefits from Vietnam's forests.

2.10.3 Finding Space for Social Forestry within Economic Development and Tenure Conflicts in Sabah

The Sabah Forestry Department (SFD), with support from the German Agency for Technical Cooperation, developed and implemented a pilot case for sustainable forest management (SFM) in Deramakot Forest Reserve in 1989 as part of a Forest Stewardship Council (FSC) certification process to address community development issues (UNDP, 2008). This pilot approach was considered as a model for SFM, and the acquired FSC certification in 1997 provided impetus for SFD to apply similar 'community forestry' approaches in other FSC forest concessions. The year also marked a significant shift of SFD's core policy approach from industrial logging towards adoption of a SFM strategy and social forestry was institutionalized within this. Under the SFM License Agreement (SFMLA), if there are communities living in or around the forest concession boundaries, the license holders are required to verify community benefits, provide dispute resolution systems and set aside land for community forestry projects within the forest concessions (Toh & Grace, 2005). While acknowledged as an improvement to previous conflicts between industrial logging and local communities, this requirement effectively renders social forestry as a legal and technical obligation for the success of the SFMLA concession. Toh and Grace (2005) argues that it can also be a mechanism for controlling community land use within the forest reserves – a territorialization process where forest managers assign strict boundaries for agriculture plots where only approved tree and cash crops can be cultivated. Persistence of colonial legislation for economic growth. Conflicts around forest access and rights stem from two key colonial laws: the Sabah Land Ordinance (SLO) of 1930, which recognizes Native Customary Land Rights and includes provisions for the granting of Native Titles for indigenous individuals and households and Communal Titles for shared access, and the Ladang Act of 1913 which was meant to control shifting cultivation. Both pieces of legislation signalled the North Borneo Chartered Company's (NBCC) intention to settle Native land rights and identify "idle" lands suitable for plantation agriculture (Doolittle, 2007; Bernard & Bissonnette, 2011; Majid Cooke & Toh, 2012). The issues of Native rights proved much more difficult for the NBCC however, and this tangled legacy of unresolved rights has led to anxieties, contestation and multiple overlapping claims to forested lands today. Applications for individual Native Titles often take many years to be resolved while logging and plantation licenses are processed quickly6. During the peak of palm oil plantation expansion in 2009, there was a reported backlog of 285,000 Native Title applications (Majid Cooke & Toh, 2012). The complexities passed on from colonial rule have paved the way for territorialization processes of zoning and the demarcation of land and forests for the long-term state objectives of land commodification and centralized control of natural resources (Doolittle, 2007;Lunkapis, 2013; Majid Cooke, 2013). In 2010, the Sabah government stopped issuing Native Titles to prioritize communal titles (SUHAKAM, 2013). In total, 96 communal titles totalling 61,6201 ha were issued to 13,789 recipients from 271 villages in 15 districts in Sabah (Borneo Post, 2012). The shift to communal titling was built on familiar discourses of low productivity, idle land and poor decisions of natives (Majid Cooke, 2012) to push forward the State's strategy for rural development and poverty alleviation (SUHAKAM, 2013; Fadzilah & Toh, 2012). An amendment of the SLO in 2009 to include the phrase 'any state land planned by the government for the natives of Sabah' made it 'possible for the creation of communal titles in cases where the government plans to develop the land, albeit for the benefit of natives' (Lunkapis, 2013). The communal title, issued through the Fast Track Land Alienation Program, requires the participants to sign an agreement to transfer their rights to state agents for creation of joint ventures for oil palm plantations on the Native lands (Lunkapis, 2013). The communal title came with 'special conditions', including expectations that the land has to be cultivated with commercial plants or trees approved by the Director of Agriculture for the duration of tenure (Majid Cooke, 2012; SUHAKAM, 2013). Large tracts of forest land degazetted for Native communal titles were effectively converted to oil palm plantation and the process was deemed to have 'violated the rights of natives' (SUHAKAM, 2013). The rapid territorialization and de-territorialization of native communal titles legitimized by policy were intended to benefit the state (and convergence of state leaders' and private sector actors' interests) at the expense of the indigenous population. Potential for a more equitable form of social forestry and native land claims? Within this dynamic context of Native land as the frontier for development objectives, SF in Sabah occupies a fraught space. The active engagement of civil society organizations and regional ASEAN Working Group on Social Forestry (AWG SF) over the past decade has led to the establishment of SASOF with the SFD in 2016. Serving as a multi-stakeholder platform to discuss and align SF activities and to bring new perspectives and ideas to definitions and understandings of SF, SASOF was instrumental in the development of Sabah and National Social Forestry Roadmap (SFD, 2018). The roadmap is undergoing consultation and revisions, and in the latest version, SF is defined as 'the involvement of indigenous peoples and local stakeholders in forest management from the aspects of social, economic, culture and environment for sustainable livelihoods'. At the same time however, social forestry and issues of communal tenure or rights remain conspicuously absent from the Sabah Forest Policy 2018 (SFD, 2018) and policy actions only emphasize increased consultations and dialogue with local and Native peoples in forest management, preferential contracting and employment in forest industries, and development of native entrepreneurship (Ibid, p.31). With funding from UNDP, the proposals of CSOs, also SASOF members, to transfer global principles of ICCA (territories and areas conserved by indigenous peoples and local communities) to Sabah provide fresh impetus to integrate Native knowledge and governance practices within SF policies, and is considered by local communities as another mechanism to advance their claims for Native rights (Majid Cooke & Vaz, 2011).

2.11 Right to Nature (Table 2.6)

'Sustainability' as a concept should guide us to improve the health of the biosphere and our relationship with it, rather than prop up current, harmful consumption and production practices. The Expert Report on Earth Jurisprudence submitted to you describes 'sustainable' behaviour as being "reconnect[ed] with Nature's processes." The Report emphasizes that nature can no longer be treated as a commodity – it is our relation, and we must evolve law and policy to reflect that fact. The concept of rights helps us understand this point more clearly. The 2030 SDG Agenda seeks to 'realize the human rights of all,' and adds that the SDGs are 'grounded in' the Universal Declaration of Human Rights. This is important, but we also must ground the SDGs in nature's rights if we are to protect human rights in practice. For example, the U.N. has acknowledged that the human right to water is 'a prerequisite for the realization of other human rights.' But how can we say we have a right to something, like a waterway, if it does not itself have a right to exist and thrive? The Earth Jurisprudence Report recommends that we fix this gap by recognizing nature's rights in law and practice, just as we recognize human rights and the rights of indigenous peoples. Just as law and policy grounded in human rights sets a higher standard for action, so too does law and policy grounded in nature's rights better protect the well-being of ecosystems and species. Defining 'sustainability' in terms of nature's right to thrive is significantly different from what we do now. We treat nature as a commodity for our economic system first, with only limited safeguards to slow nature's degradation. This approach is failing, as species and habitats rapidly disappear and the global temperature continues to warm. We must set our sights higher. The 2030 Agenda makes room for nature's rights, calling for a 'healthy environment' and declaring a vision of a world 'where all life can thrive.' The structure of our laws, policies, and scientific inquiry must support this core vision of sustainability.

Putting profit before people and nature is unsustainable development. We must redefine what we expect sustainable business and finance to look like, and move

Table 2.6 Sustainable development goals and associated human rights related towards right to nature

Sustainable development goals		Related human rights
12. Responsible consumption and production	Ensure sustainable consumption and production patterns Targets include achieving sustainable management and efficient use of natural resources; improving waste management; promoting sustainable public procurement; ensuring access to information; and building capacity for sustainable development.	(i) Right to health including the right to safe, clean, healthy and sustainable environment (ii) Right to adequate food and the right to safe drinking water (iii) Right of all peoples to freely dispose of their natural resources
13. Climate action	Take urgent action to combat climate change and its impacts targets include strengthening resilience and adaptation to climate change and natural disasters, including in marginalised communities; implementation of the Green Climate fund.	(i) Right to health (ii) Right to adequate food & right to safe drinking water (iii) Right of all peoples to freely dispose of their natural wealth and resources
14. Life below water	Conserve and sustainably use the oceans, seas and marine resources for sustainable development targets include reducing marine pollution; conserving costal ecosystems, costal marine areas and fish stock; securing market access for small scale fishers; protection of marine biodiversity.	(i) Right to health (ii) Right to adequate food & right to safe drinking water (iii) Right of all peoples to freely dispose of their natural wealth and resources
15. Life on land	Protect, restore and promote sustainable use of terrestrial ecosystems, sustainably manage forests, combat desertification, and halt and reverse land degradation and halt biodiversity loss targets include the sustainable management of freshwater, mountain ecosystems and forests; combatting desertification; halting biodiversity loss; combatting poaching and trafficking of protected species.	(i) Right to health (ii) Right to adequate food & right to safe drinking water (iii) Right of all peoples to freely dispose of their natural wealth and resources

away from a primary profit objective and toward expanding benefits for Earth society, which includes all life. What does this mean in practice? One conclusion is that we need to replace gross domestic product as a desirable measure of sustainable progress. The Earth Jurisprudence Report observes that 'blind adherence to economic growth as a measure of well–being has resulted in increasing harm to the planet and to all of us.' GDP growth does not translate to sustainable development; it includes all development, both good and tragic. Earlier this year, the World Economic Forum itself criticized such economic growth as failing to address the needs of the political economy. Sustainable development calls on us to choose our actions in light of their broader impacts, rather than continue to focus on short-term profit. Fortunately, this economic evolution is already beginning. More businesses are measuring success not just by profits and economic growth, but also by the measurable social and ecological benefits they generate. More investors are demanding investment products that offer both financial returns and strong social and ecological returns. This expanded focus allows for improved consideration and management of risk, and so can result in better financial performances over time than conventional instruments.

One such example is the increasing number of investments related to mitigating climate change, to help the world stay below the 2 °C threshold. The social and ecological impacts of climate change are potentially catastrophic and irreversible. Financial investments that recognize and mitigate such risks, such as investments in the low carbon economy, will ensure sustainability far more effectively than those that merely increase GDP. It is with climate change in mind that the World Economic Forum also identified an 'urgent' need for 'long-term thinking' in our economic system. To be sustainable in fact, development must be conceived and implemented in light of the magnitude of the long-term risks of climate change and other global stressors. Our sustainable development objective, however, is not merely occasional success. We must create instead a self-perpetuating system, designed to generate development and investment behaviour that regularly advances net positive impacts for human and ecological communities. We need to think beyond current investment and economic practices to do this. For instance, we should consistently place higher costs on economic behaviours that cause injury, such as fossil fuels, and reward with higher returns those economic choices that measurably benefit people and planet. Tying our economic, business, and finance priorities to progress in safeguarding nature's rights and human rights will help us measure our success and keep us on track. We also must evolve our characterization of nature as an economic commodity. Our relationship with nature is fundamentally familial and local. As the Earth Jurisprudence Report recommend, development must reconnect with and respect nature's processes, rather than destroy those processes in the name of profit. We have little time to course-correct before the impacts of climate change, species extinctions, and habitat destruction create tipping points that irreversibly injure all life. The key is to focus on the 2030 Agenda's core goal: 'a world where all life can thrive.' This includes the natural world, as we are inseparable from it. Our core goal for 2030 is not sustainable development. Our goal instead is a thriving world; sustainable development is just a tool. Nature's rights provide a necessary foundation from which our governance systems can build truly sustainable development practices that will help us achieve our goal, including finance and business systems designed to seek out and maximize ecological and social gains. Earth is the source of the laws that govern our lives in every respect. Our human laws must heed Earth's laws. Recognition of nature's rights in local, national, and international law and policy will help keep us on that path (Sheehan, 2017).

Finally, it will be vital to understand the ecologies of coexistence, or 'reconciliation ecology'. Decisions to evict people, or restrict their access to resources, should be governed by pragmatic ecological considerations rather than ideals of wilderness. Yet for some conservationists it is the wilderness ideal that really matters.

Strictly protected areas containing people are anathema. In many countries national parks are envisaged as places where rural livelihoods do not belong. In these contexts, people in parks are a category error. But it is an invidious one. It fosters antagonisms: there are several cases of animals and forests being deliberately destroyed to avert planned protection. The category error also threatens degradation because displaced rural livelihoods can cause considerable environmental problems. Finally, it ignores local conservation initiatives; rural groups all over the world set aside places in which they restrict their own resource use and residence. The ecologies and social impacts of coexistence are most important because this is an issue that extends far beyond the boundaries of protected areas. Six percent of the world is, after all, not very much from the perspective of conservation on a global scale. The ultimate challenge facing conservationists today is not only to reconcile errors of the past but also to determine how to shape human interactions with nature in landscapes of which people are a part (Brockington et al., 2006).

References

Berkes, F. (2007). Community-based conservation in a globalized world. *Proceedings of the National Academy of Sciences, 104*(39), 15188–15193. https://doi.org/10.1073/pnas.0702098104

Bernard, S., & Bissonnette, J. F. (2011). Oil palm plantations in Sabah: Agricultural expansion for whom. In J. F. Bissonnette, S. Bernard, & R. DeKoninck (Eds.), *Borneo transformed: Agricultural expansion on the Southeast Asian frontier* (pp. 126–151). Singapore.

Bong, I. W., Moeliono, M., Wong, G. Y., & Brockhaus, M. (2019). What is success? Gaps and trade-offs in assessing the performance of traditional social forestry systems in Indonesia. *Forest and Society, 3*(1), 1. https://doi.org/10.24259/fs.v3i1.5184

Borneo Post. (2012) *Native land issues dominate second day of hearing*, published, https://www.theborneopost.com/2012/06/06/native-land-issues-dominate-secondday-of-hearing/. Accessed on 31 Aug 2021.

Brockington, D., Igoe, J., & Schmidt-Soltau, K. A. I. (2006). Conservation, human rights, and poverty reduction. *Conservation Biology, 20*(1), 250–252. https://doi.org/10.1111/j.1523-1739.2006.00335.x

Clement, F., & Amezaga, J. M. (2008). Linking reforestation policies with land use change in northern Vietnam: Why local factors matter. *Geoforum, 39*(1), 265–277. https://doi.org/10.1016/j.geoforum.2007.05.008

Conservation Initiative on Human Rights. (2014). *Human rights in conservation: Progress since Durban* (pp. 1–13). White Paper.

Cummins, A., & Yamaji, E. (2019). To see invisible rights: Quantifying Araman informal tenure and its immediate relationship with social forestry in Central Java, Indonesia. *Forest and Society, 3*(2), 193–201. https://doi.org/10.24259/fs.v3i2.6289

Dang, T. K. P., Turnhout, E., & Arts, B. (2012). Changing forestry discourses in Vietnam in the past 20 years. *Forest Policy and Economics, 25*, 31–41. https://doi.org/10.1016/j.forpol.2012.07.011

De Royer, S., Van Noordwijk, M., & Roshetko, J. M. (2018). Does community-based forest management in Indonesia devolve social justice or social costs? *International Forestry Review, 20*(2), 167–180. https://doi.org/10.1505/146554818823767609

Doolittle, A. A. (2007). Native land tenure, conservation, and development in a pseudo-democracy: Sabah, Malaysia. *The Journal of Peasant Studies, 34*(3–4), 474–497. https://doi.org/10.1080/03066150701802793

Erbaugh, J. T. (2019). Responsibilization and social forestry in Indonesia. *Forest Policy and Economics, 109*, 102019. https://doi.org/10.1016/j.forpol.2019.102019

Fadzilah, A. M., & Toh, M. (2012). SUHAKAM National Inquiry into the land rights of indigenous peoples and access to customary lands: A question of rights in contemporary Sabah.

Fisher, M. R., Moeliono, M., Mulyana, A., Yuliani, E. L., Adriadi, A., Judda, J., & Sahide, M. A. K. (2018). Assessing the new social forestry project in Indonesia: Recognition, livelihood and conservation? *International Forestry Review, 20*(3), 346–361. https://doi.org/10.1505/146554818824063014

Forest Survey of India (FSI). (2003). *State of forest report 2001.* Ministry of Environment and Forest.

KSP (Kantor Staf Presiden), (2016). Strategi Pencapaian Target Program PrioritasReformaAgraria. https://hanibalhamidi.files.wordpress.com/2016/09/naskah-stranas-pelaksanaan-reforma-agraria-ksp.pdf. Accessed on 10 Sept 2021.

Lasco, R. D., & Pulhin, J. M. (2006). Environmental impacts of community-based forest management in the Philippines. *International Journal of Environment and Sustainable Development, 5*(1), 46. https://doi.org/10.1504/ijesd.2006.008682

Lunkapis, G. J. (2013). Confusion over land rights and development opportunities through communal titles in Sabah, Malaysia. *Asia Pacific Viewpoint, 54*(2), 198–205. https://doi.org/10.1111/apv.12019

Majid Cooke, F. (2012). In the name of poverty alleviation: Experiments with oil palm smallholders and customary land in Sabah, Malaysia. *Asia Pacific Viewpoint, 53*(3), 240–253. https://doi.org/10.1111/j.1467-8373.2012.01490.x

Majid Cooke, F. (2013). Constructing rights: Indigenous peoples at the public hearings of the National Inquiry into customary rights to land in Sabah, Malaysia. *Journal of Social Issues in Southeast Asia, 28*(3), 512. https://doi.org/10.1355/sj28-3e

Majid Cooke, F., & Toh, S. M. (2012). *Indigenous peoples and access to customary lands: A question of rights in contemporary Sabah.* Report submitted to the Human Rights Commission of Malaysia.

Majid Cooke, F., & Vaz, J. (2011). *A review of indigenous and community-conserved areas in Sabah.* Global Diversity Foundation for the Sabah Biodiversity Centre, Bornean Biodiversity and Ecosystem Conservation Project II.

MARD (Ministry of Agriculture and Rural Development Vietnam). (2017). *The new Forestry Law will impact on 16, 2 million hectare of forest,* https://www.mard.gov.vn/Pages/bo-nnptnt-luat-lam-nghiep-se-tac-dong-den-16-2-trieu-ha-dat-rung.aspx

MOEF. (2003). *JFM Cell.* Ministry of Environment and Forests.

MOEF (Ministry of Environment and Forestry). (2018). State of Indonesia's Forest 2018, Jakarta.

Moeliono, M., Thuy, P. T., Bong, I. W., Wong, G. Y., & Brockhaus, M. (2017). Social forestry-why and for whom? A comparison of policies in Vietnam and Indonesia. *Forestry and Society, 78–97.* https://doi.org/10.24259/FS.V1I2.2484

Murali, K., Murthy, I. K., & Ravindranath, N. (2002). Joint Forest Management in India and its Ecological Impacts. *Environmental Management and Health, 13*(5), 512–528. https://doi.org/10.1108/09566160210441807

Myers, R., Intarini, D., Sirait, M. T., & Maryudi, A. (2017). Claiming the forest: Inclusions and exclusions under Indonesia's 'new'forest policies on customary forests. *Land Use Policy, 66,* 205–213. https://doi.org/10.1016/j.landusepol.2017.04.039

Ngo, T. D., & Mahdi. (2017). Targeting deforestation through local Forest governance in Indonesia and Vietnam. *Redefining Diversity & Dynamics of Natural Resources Management in Asia, 1,* 273–288. https://doi.org/10.1016/b978-0-12-805454-3.00014-1

Pham, T. T., Ha, H. M., & Campbell, B. M. (2008). Pro-poor payments for environmental services: Challenges for the government and administrative agencies in Vietnam. *Public Administration and Development, 28*(5), 363–373. https://doi.org/10.1002/pad.513

Pham, T. T., Moeliono, M., Nguyen, T. H., Nguyen, H. T., & Vu, T. H. (2012). The context of REDD+ in Vietnam: Drivers, agents and institutions. *CIFOR. Occasional Paper, 75.* https://www.cabdirect.org/cabdirect/abstract/20123320968

Pham, T. T., Bennet, K., Vu, T. P., Brunner, J., Le, N. D., & Nguyen, D. T. (2013). *Payments for forest environmental services in Vietnam: From policy to practice. Occasional paper 93*. CIFOR.
Pham, T. T., Di Gregorio, M., Carmenta, R., Brockhaus, M., & Le, D. N. (2014). The REDD+ policy arena in Vietnam: Participation of policy actors. *Ecology and Society, 19*(2). https://doi.org/10.5751/es-06389-190222
Poffenberger, M. (2006). People in the forest: Community forestry experiences from Southeast Asia. *International Journal of Environment and Sustainable Development, 5*(1), 57–69. https://doi.org/10.1504/IJESD.2006.008683
Ravindranath, N. H., Murali, K. S., Murthy, I. K., Sudha, P., Palit, S., & Malhotra, K. C. (2000). Summary and conclusions. In N. H. Ravindranath, K. S. Murali, & K. C. Malhotra (Eds.), *Joint Forest management and community forestry in India: An ecological andInstitutional assessment* (pp. 279–318). Oxford and IBH.
Ravindranath, N. H., Murali, K. S., & Sudha, P. (2006). Community forestry initiatives in Southeast Asia: A review of ecological impacts. *International Journal of Environment and Sustainable Development, 5*(1), 1–11. https://doi.org/10.1504/ijesd.2006.008678
RECOFTC. (2014). *Current status of social forestry in climate change mitigation and adaptation in the ASEAN region: Situational analysis 2013*. RECOTC.
RECOFTC. (2018). *Resilient forest landscapes: Empowered communities, strengthened institutions and shared prosperity (strategic plan 2018–2023)*. TheCenter for People and Forests.
Resosudarmo, I. A. P., Tacconi, L., Sloan, S., Hamdani, F. A. U., Alviya, I., & Muttaqin, M. Z. (2019). Indonesia's land reform: Implications for local livelihoods and climate change. *Forest Policy and Economics, 108*, 101903. https://doi.org/10.1016/j.forpol.2019.04.007
Sahide, M. A. K., & Giessen, L. (2015). The fragmented land use administration in Indonesia – Analysing bureaucratic responsibilities influencing tropical rainforest transformation systems. *Land Use Policy, 43*, 96–110. https://doi.org/10.1016/j.landusepol.2014.11.005
Sahide, M. A. K., Fisher, M. R., Erbaugh, J. T., Intarini, D., Dharmiasih, W., Makmur, M., & Maryudi, A. (2020). The boom of social forestry policy and the bust of social forests in Indonesia: Developing and applying an access-exclusion framework to assess policy outcomes. *Forest Policy and Economics, 120*, 102290. https://doi.org/10.1016/j.forpol.2020.102290
SFD (Sabah Forest Department). (2018). *Sabah forest policy 2018*. Sandakan.
Sheehan, L. (2017). *The role of nature's rights in achieving the sustainable development goals. Interactive dialogue of the general assembly on harmony with nature*. Trusteeship Council, UN Secretariat Building.
Sikor, T., & Nguyen, T. (2011). *Realizing Forest Rights in Vietnam: Addressing Issues in Community Forest Management*. http://dlc.dlib.indiana.edu/dlc/handle/10535/7810
Springer, J., Campese, J., & Painter, M. (2011). Conservation and human rights: Key issues and contexts. In *Scoping Paper for the Conservation Initiative on Human Rights*. World Wildlife Fund. https://www.ohchr.org/Documents/Issues/Environment/ImplementationReport/IUCN3.pdf
SUHAKAM. (2013). *Report of the national inquiry into the land rights of indigenous peoples*. Human Rights Commission of Malaysia. https://nhri.ohchr
Toh, S. M., & Grace, K. T. (2005). Case study: Sabah Forest ownership. *Food and Agriculture Organization of the United Nations*. http://www.fao.org/forestry/10515-0462aee542d13dc983cd2bb1d09c073fa.pdf
UNDP. (2008). *Malaysia Sustainable Community Forest Management in Sabah*. https://theredddesk.org/sites/default/files/resources/pdf/2012/forest_mgmt_final.pdf. Accessed on 31 Aug 2021.
Vietforest. (2018). Rationales to enhance community forestry. Available at: http://vietfores.org.vn/tin-tuc/mot-ly-chinh-de-phat-trien-lam-nghiep-cong-dong-o-viet-nam/. Accessed on 12 Sept 2021.
Wong, G. Y., Moeliono, M., Bong, I. W., Pham, T. T., Sahide, M. A., Naito, D., & Brockhaus, M. (2020). Social forestry in Southeast Asia: Evolving interests, discourses and the many notions of equity. *Geoforum, 117*, 246–258. https://doi.org/10.1016/j.geoforum.2020.10.010

Chapter 3
Biorights of Commons

Dipayan Dey

3.1 Introduction

Biorights of commons is an innovative conservation paradigm for protecting valuable and vulnerable ecosystems that can be implemented through a community-led intervention in any socio-ecological settings, especially in global south. While most of the valuable ecosystems, for example riverine floodplains, wetlands, mangroves and coastal estuaries, forests and mountain pastures etc. support marginal and indigenous communities with their goods and services to sustain livelihood and ensure food security as well, the ecosystems themselves are subtle and susceptible to climate change impacts. Enriched resources of nature have ever allured communities to settle around, especially those who are surviving at the bottom of economic pyramids, and leverage the goods and services of nature in sustaining their needs to thrive better. From the initial stages of settlements, till the carrying capacity of the habitat could accommodate, the ecosystem remains resilient to this anthropogenic pressure. As the pressure of population rides on them, they become more and more fragile and in no time experience a sharp fall in ecosystem services, primary productivity and resilience. Impacts of climate change draw this threshold of resilience even nearer compelling it to succumb much earlier to this pressure, than expected. This in turn, furthers poverty hike due to resource crunches and finally leads to habitat destruction and loss of biodiversity owing to overexploitation of nature. Dependent communities inadvertently keep extracting resources from nature for survival, as they hardly understand the vicious nexus and slowly get entrapped in the poverty-risk labyrinth.

A scenario analysis in a periurban wetland near developing cities of eastern India like Kolkata and Guwahati will help in understanding it better. These wetlands support captive fisheries, which is a major livelihood for the marginal fisher's

D. Dey (✉)
South Asian Forum for Environment, Kolkata, West Bengal, India

D. Dey, J. Roy (eds.), *Biorights*, Studies in Ecological Economics 7,
https://doi.org/10.1007/978-3-030-91503-2_3

communities staying around it and as well the wetland ecosystem extends the service of absorbing city's effluents and overflowing waters, thereby saving the city from flood situation. Whereas, in lean winter months the fish production in these wetlands is at its minimum and fisher's communities have no other economic alternatives to sustain livelihood, at the same time, this is a suitable period for undertaking conservation efforts in the wetland habitats, like desilting, pond treatments and habitat restoration or conserving local biodiversity etc. During this period mostly the community members move out in search of jobs and temporarily migrate as casual labourers and there is no one to conserve the wetland habitat. The city's effluents and overflows keep coming in the wetlands, furthering eutrophication, deterioration of water quality and loss of aquatic flora and fauna. Subsequently production drops and fishers lean back to intensive aquafarming promoting usages of synthetic feed and fertilizers. If these members from the communities are to be engaged in conservation and habitat restoration during this lean period, then the economic opportunities they would have to forgo, by abandoning their temporary migration in search of informal jobs in urban sectors as casual labourers, must be compensated. However, in actuals there is no such compensatory instrument in place, either societal or legal, and finally with the fall in productivity and per beneficiary pay-outs or income, the community beneficiaries land up in urban informal sectors being victims of social abuse and inequity. Similar case studies are available in rural agrarian sectors or coastal floodplains that explain the same circularity of distress in the absence of a sustainable conservation approach. In deltaic villages of marginal fishers of coastal Bangladesh or eastern India, women who collect fry shrimps in estuaries during the ebb tide, also collects many fish fries, spawns and aquatic planktons those are either destroyed or disposed as valueless by-catch, causing severe losses in coastal biodiversity. This has its spell on the coastal ecosystem making it more vulnerable to climate impacts like storm surge, inundation and flooding. To prevent the same, the women fishers need to be compensated with alternative livelihood opportunities, failing which the rights of indigenous people to have access to nature's goods may be put to question. In other words, the dependence on nature has to be transformed to a self-paying conservation effort. These are discussed in more details in the later part of this publication.

In such precarious condition, any ex-situ or top-down conservation approach with development priorities may trigger uneven challenges of displacement, violation of community rights, agitations and may also overcast with adverse impacts on the ecological system, dilapidating significant natural processes like vegetation dynamics and bio-geochemical cycles. On the contrary, community efforts in conserving the habitat ecology might cost a considerable price for the opportunities lost by these communities in accepting changes for conservation measures. However, it is equally important to make all such conservation measures socially inclusive and equitable, since the local inhabitants would only have the actual onus of preserving the ecosystem and for ensuring sustainability, collective action and participatory decision making by the community beneficiaries and stakeholders will be crucial. In the milieu of climate change, biorights model tries to protect areas of global socio-ecological importance by engaging local communities in conservation activities and sustainable practices so that the pressure of anthropocene on the ecosystem is lessened. It equally attempts to compensate the marginal commons, for their opportunities lost in protecting nature, who live near these nature areas and are dependent on its goods and services for cash generating activities. The hypothesis thus implies that establishment of a win-win situation to ensure sustainability reciprocity and equity, as well as disjoining the negative link between poverty, people and nature, can be accomplished through biorights of commons. In the context of sustainable development, biorights can be a very attractive approach at community-ecosystem interface, as the economic compensation for lost opportunities can be regained by transforming PES to a fiscal module for the communities through cash generation activities.

In an attempt to highlight the economic features of biorights which is integral to its concept and to elucidate it as a financial tool for both poverty alleviation and environmental conservation, analytical scenario planning studies have been conducted in various socio-ecological sites and socio-political settings in south and south east Asia. It is interesting to note that the economic value of nature is only marginally present in the market as it has not been recognised so far by the global development leaders or policy planners and only market pricing of nature's goods could be acknowledged with our current economic system. Therefore, in the absence of true values of nature that consists of both direct and indirect values, the costs of nature conservation are minimally visible in the market. Only the most obvious benefits, such as tourism revenues or renewable energy sources are usually accounted for in the market. This becomes a predicament in the current context to make an unbiased benefit to cost analysis of investments in existing nature reserves. Therefore, it is equally difficult to equate the cost of nature conserved or cost of habitats restored against which the contributions of commons in conserving or restoring nature could be paralleled.

Studies revealed that in developing countries biorights could directly contribute to conservation by compensating the opportunity costs of local people for not degrading the natural environment. The global average cost needed to compensate the same, lies in the range of US$13.65 ha-1 yr-1, which is assumably nominal (Dey, 2010). Since biorights does hold elements those have potentials to raise this amount by liquidating payments from ecosystem services, feasibility in deploying the biorights model for community-based conservation needs to be assessed. The model can also facilitate direct payments to compensate for poverty related costs to guarantee long term sustainability by abating temporary migration and capitalize on community engagements. In one way, biorights incentivize the efforts of communities for measures in environmental conservation as a 'common good' for people by paying equitably. However, to ensure equity and accountability, distribution of payments may be done to joint liability groups in the community and not to specific community beneficiaries on the basis of their individual efforts in nature conservation. It is evident that all collective practices and yields will leverage from the 'economy of scale', which can't be ensured with singularity of action. Thus, there has to be a critical number of beneficiaries to be engaged in conservation activities. This 'critical number' essentially depends upon the volume of conservation work, status of the ecosystem health and number of people that the ecosystem can sustain. Thus, biorights would also be underpinning shared responsibilities and shared values to better mitigate the community risk in the climate milieu. At the same time, if the climate risk of vulnerable communities is dissipated and mitigated by restoring the habitat and sustainably intensifying the ecosystem services, then it would be a compensation in itself and will be incentivizing their efforts in conservation. Herein the fiscal compensation may also be incorporated as an insurance premium for risk coverage. In the design section, we would elaborately discuss it.

3.2 Genesis of Biorights

While settling around any ecosystem, preferences of marginal communities uniquely depends on a couple of things, of which the most important being the primary productivity potentials of the habitat and its wide array of ecosystem services. For instance, all farmers, settling in the coastal and riverine floodplains as rural agrarian community, benefits from the flood borne alluvial silt that enriches the soil and ensures good agricultural production, whereas fishers leverage the floodwaters and wetlands for both captive and capture fisheries. In the periurban settings, marginal communities in shanty towns are usually found to be flocking around a lake or waterbody, as its ecosystem services are essential for their daily chores, while captive fisheries can also be practiced in the waterbody, as well. However, there are much wider variations in community-ecosystem associations and those are mostly place-based. For example, indigenous highlanders prefer high altitude pastures for raring and grazing of their livestock that sustains their livelihood. The services of the ranging ecosystem are therefore equally important with its productivity potentials to feed the livestock grazing on it. Similarly, many forest communities largely depend on the non-timber forest products for livelihood and the services as well, that they retract from the forests or the species having medicinal values that are essential for sustaining their life locally. These preferences gradually build up their dependence on ecosystem goods and services, available free in nature.

Dependence on ecosystem goods and services in indigenous and rural communities are far more intense than communities in periurban settings as the alternatives or substitutes for the goods and services are almost unavailable in rural, whereas in periurban or urban settings there can be some stand-ins. This has two implications in the rural of which one is definitely rational consumption and wise use. There are many instances of refurbishing the species often used by the indigenous communities, preservation of biodiversity and restricted usages as an inherent social code of conduct in the communities. In most of the indigenous communities of north-eastern India, these practices are prevalent. For example, the Himalayan highlanders like *Aapatani, Nyishi, Mising* etc. design their crop cycles in a way that it ensures the conservation of the local agro-biodiversity. Likewise, The *Sarchop* tribes of eastern Himalayas in Bhutan, who extract essential aromatic oil from natural vegetation of *Cymbopogon flexuosus* (Lemon grass), maintain intermittent fallow lands and put restricted fires in patches of hill slopes for facilitating pyrecological growth of the grass. Similarly, there are instances of releasing fish fries and fingerlings in local community waterbodies from rivers to maintain the brood stock and aquatic species diversity or regulations like 'No-fishing' period, 'forest closure' period etc. Though there exist opportunity costs in these cases too, but this is willingly foregone by the communities to assure some sort of societal security for food, fodder or even community health, which helps in accomplishing community level sustainability, as well. The other implication is that the indigenous and rural communities often incorporate nature's goods or essentially used species in their ethno-cultural rituals that prompts a social practice of conservation of the same. In conserving the goods or species they also engage in protecting its habitat. Gradually, these rituals get inherited across generations and develop a legacy amongst the communities, ensuring community-based conservation practices. Ethnobiological usage of more than 140 forest species by the transhumant agro-pastoralists in eastern Bhutan is examples thereto (Dey & Pankaj, 2003).

It has been observed that the traditional practices in the rural indigenous communities are better preserved, if the communities are geographically or spatially distanced from urbanized sites or developmental activities. There is always a trade-off between human habit and habitat. The indigenous communities or rural people usually customize their habits based on the available resources, especially natural resources in their habitat. But when urbanization and developmental projects infringe their way, the human habits get altered to fetch better amenities, assured returns and secured livelihood. The agrarian economy locally becomes market driven and the habit of conserving the local tradition slowly fades out, since alternatives starts appearing in the local markets. In eastern Bhutan, for instance, buckwheat, a local crop that grows in uphill pastures and used for making noodles, were cultivated with traditional no-tillage farming, so long packeted fast foods appeared in local shops. Since then, buckwheat lost its local demand and the fields were converted to potato farms to be grown for chips making industries. Intensive farming and tillage were found to be the reason for deteriorating soil functions; an important ecosystem service that was availed till the substitutes of Buckwheat was available. This change of practice in socio-cultural transition came as a spin-off of economic development and this has often been debated on the context of price and values of economic development for indigenous folks. From here, we take this discussion ahead to find the genesis of biorights in the given context explaining how the perception of biorights can be seeded as a core sustainability component amidst both conservation and development.

Whenever a community or a group of inhabitants use natural goods and services of their habitat and protect it as well for their own interest, they earn their natural rights on it. This is socially accepted and can also be legally established on the pretext of human rights of the inhabitants. As discussed earlier, those chosen habitats are often found to be of ecological significance owing to its rich bio-resources and services, whereas the inhabitants therein continually strive for better lifestyle and amenities, at urban equivalence, as a right to survive. Though such types of community rights are brisk, the challenges that prevail with the dilemma of sustaining conservation and promoting development in such areas of ecological significance is to conserve or restore the ecosystem health without denying the rights to development of the inhabitants. With the advent of developmental measures, as the market demand for commodities sharply increase, pressure on natural resources too, intensifies at large. In this situation, community beneficiaries prefer developmental packages over conservation programmes for obvious reasons and whenever the concerns of environmental degradation are brought forth, demands for developments, human rights issues and mass agitations politicize the social ambience. In such hegemony of conflicts, biorights tries to pledge a logical solution leveraging the bondages of communities with the goods and services of their natural habitat through the social rituals and cultural practices g developed over time. It motivates people in conserving nature by restoring their habitat, as well refurbish their socio-cultural identities, which in turn engage them in stewarding conservation efforts for their ecosystem. Further, it takes advantage of the situation by collectively compensating for the development opportunities lost in conserving their socio-cultural or traditional legacy with the advent of developmental programmes, thereby proclaiming their right to equal and reciprocal partnership in the programme. The compensation comes through Payments from Environmental Services (PES) in the forms of financial inclusion and collective risk mitigation measures. Eventually, this strengthens the local institutions and incentivizes their efforts in conservation. More they protect the natural resources and services more it gets enriched and secure better returns for them, whereas equitable partnership in development processes empower the communities to establish their natural rights.

There are contrasting differences in the transformative processes of biorights between rural and periurban settings that owe its occurrence to the presence or absence of land rights of commons. In periurban space, usually the marginal communities living in slums or shanty towns provisionally prefer to settle on vested government lands or around waterbodies wherein they do not pay rentals and neither enjoy land ownership, but are allowed to use the resources and services of the habitat and its ecosystem. The activities for income generation from it are, therefore, found to be a collective endeavour of the communities, who either distribute the returns equitably or have contractual agreements for benefit sharing among actors, as may be suitable. These communities usually have a floating populace and therefore the heterogeneity of user rights and plurality of inhabitancy both becomes impediments of their natural rights, though they deserve equitable access to better facilities and urban amenities on humane considerations. In this case, biorights acts as a catalyst of change and considers them as partners of conservation, who take the onus of restoring habitat health. Here too, they are equitably compensated for the opportunities lost by them in pursuing the conservation program by transforming the payments from ecosystem services that they sustainably intensified through ecosystem-based adaptive practices. This incentivizes their entrepreneurial efforts in accruing economic benefits. Biorights facilitates in building suitable capacities in them for leveraging alternative economic opportunities in leu of the opportunities lost, whereas compensations help in risk mitigation.

3.3 Biorights: Design Thinking Approach

The human percept of thinking about a model or paradigm follows a patterned thought process that is often build on existing knowledge, predefined acuities and concepts those are already established and reviewed. Though such practices are less prone to risks of model failure but may be rhetoric and repetitive in nature. It is common in occurrence because it helps us in applying same action, similar knowledge or familiar situations more quickly to resolve a problem. Such patterned thinking are often called schemas, which are sets of organized information, links, relations, stimulated actions or even stimuli that the mind has encountered before and recall from memory on a specific situation or known environment. Thus, a single schema may contain vast amount of information and therefore may also prevent us from exploring new data sets or perceptible characters. It may also obstruct a more fitting impression of the situation or prevent us from seeing a problem in a way that will enable a new problem-solving strategy. A much better way of understanding something intangible or abstract, such as biorights and ecosystem services, is by understanding what it is not. This would need out of box thinking designs, foe innovative problem solving. This is one reason, biorights has a design thinking approach to make it more adaptive over space and time.

It is interesting to note that design thinking uses the same information stock but it helps us to systematically extract, use and apply that information facilitating adaptive learning. Its important especially for human centric techniques to solve problems, in a more creative and innovative way. Thus, biorights too, has a core component based on design thinking and it is important to explore it in understanding the model. Being a neo-economic paradigm for conservation, biorights happens to be an iterative process that understand users, challenge assumptions and redefine problems in an attempt of identifying strategic solutions and alternatives that might not be instantly apparent with our initial level of understanding. At the same time, it aspires to provide a solution-based outcome achievable with hands-on methods through prototyping and testing. Biorights can thus be re-framed with design thinking approach and be made more empathic with a human-centric approach so as to tackle more ill-defined and unknown problems more innovatively at community-ecosystem interface.

We will discuss the Design Thinking in Biorights in three separate perspectives, viz. attributes, model and efficacy. While the attributes will describe the unique features of the design in biorights, the Model will explain its functionality and based on these we can assess the efficacy of the model in resolving related problems in varying socio-ecological milieu. The attributes of biorights are

(a) *Circularity:* It is an important attribute of the design facilitating activities and outcomes to circulate within the paradigm, which allows the same event or activity to be an input at one point and as output in another. Such interconnective loops of actions between components of biorights make it flexible and adaptive.

(b) *Feedback Inhibition*: The design has an inbuilt self-regulatory mechanism to inhibit or prevent some action or activities through a feedback mechanism. Such feedbacks are auto generated from information girdles and delink the circular processes to stop an action. This helps to prevent loss of resources and prevents logjam in multi-criteria decision support systems.
(c) *MCDA - Matrix*: Applications of biorights model across varying socio-ecological landscapes and seascapes brings in trade-offs and synergies between the sustainability factors. To make it highly resilient to changes, especially in the climate milieu, the model has been based on multicriteria decision analysis and support system. In plurality of conflicts between conservation and development multiple factors need to be explicitly evaluated in decision making. In climate vulnerable areas primary producers are interested in getting high returns while simultaneously reducing climate risks. However, the crop that have the potential of bringing high returns typically carry high risk of production in aberrative climatic regime. The potentials of ecosystem-based adaptation (EbA) to provide intended climate benefits needs to be assessed on the context of trade-offs that may arise if that EbA leads to maladaptation owing to a low biodiversity value. Thus, in building up biorights of commons, MCDA and support system becomes inevitable for ensuring equity and reciprocity in a community-based participatory model.
(d) *Valuing Nature Services*: Biorights being a neo-economic conservation paradigm aspires to compensate the opportunity costs by transforming the nature services to fiscal equivalents, pricing the intangible services and estimating their direct and indirect values are essentially integrated with the model design.
(e) *Closed Systemic Design:* Since, biorights is a place-based intervention, the design of the model assumes that it is a closed system operational in the local context and external market dynamics or major socio-ecological changes are not infringing its boundaries, unless happening in the same location or habitat. Therefore, the conservation module that the design tries to sustain and protect is socio-ecologically a part of the larger ecosystem that constitutes the habitat of the marginal communities living within.

The design is based on the general principles of thermodynamics that studies the interaction between large ensembles called the system and the surroundings. It assumes that there is a randomness of interaction, called entropy, between the system and the surroundings owing to the exchange of energy between them and this brings changes called the enthalpy. The dynamic system continually attempts to come to an equilibrium. Herein, if the socio-economic attributes of biorights like natural rights, compensations, economic opportunities, ecosystem services, etc. implied in conservation, are collectively considered to be a system then the natural ecosystem is its surrounding and the anthropogenic activities are the interactions between them. This design now tries to explain how the biorights system responds to its surrounding ecosystem in this anthropocene and explicitly defines the changes met as the outcomes. Thermodynamically, a system can evolve and quantify the state of order it is in and that can be used to further quantify the outcomes and impacts those could be mined from the same system. Therefore, the outcomes and the impacts are the indicators of change that alters with time and as well, randomness of the system and surroundings. This is schematically represented in the accompanying figure (Fig. 3.1).

Perusal of the design in the figure that enunciates biorights in a 'design thinking' approach (Fig. 3.1) reveals that there are three interactive segments or domains of biorights, namely 'activity', 'components' and 'outcomes', spatially arranged in vertical layers with diminishing randomness. The lowermost is the activity layer with maximum randomness, middle one is the layer of components that catalyse the changes and upper layer is the outcome or measurable impacts with much subdued randomness. This represents an up-hill wave of entropy that slowly dampens while percolating through the design components only to churn out a tangible change (enthalpy) as an outcome. The components of this closed system (biorights) continually attempt to accomplish a dynamic equilibrium between itself and the surroundings and that is what would mean as the sustainability or resilience of the system. The first domain of 'Activity' comprises of the human activities around the system, namely – livelihood and income generation activities in the non-farm sector that may include manufacturing, sales, services etc., – primary production like farming or fisheries as farm sector activities, – risk mitigation activities for socio-economic security like disaster preparedness, food and water security, financial inclusion etc., – developmental activities like housing and infrastructure, education, health, water and sanitation etc. there may be several other activities and all are equally important in the context of a sustainability and inclusive growth.

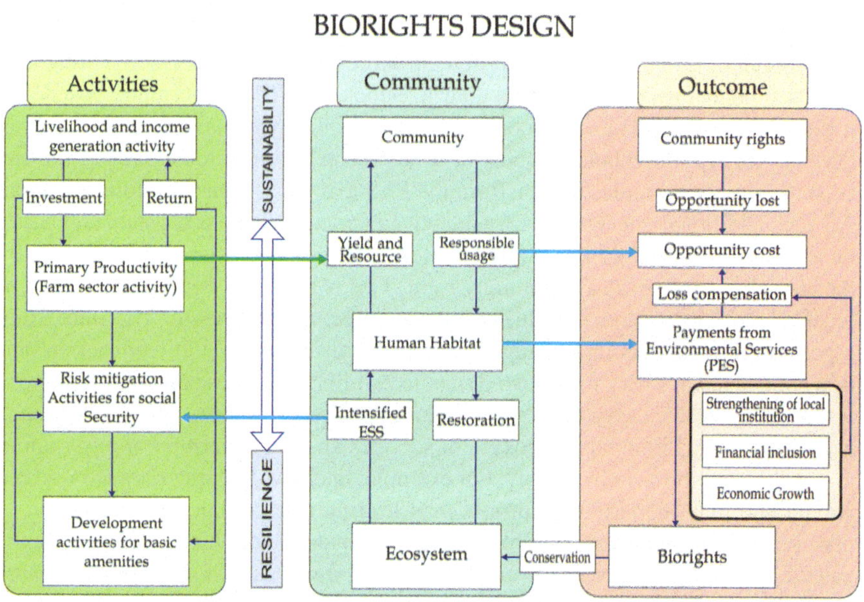

Fig. 3.1 Design thinking approach in biorights

The next segment defines the components of biorights, those make the surroundings. It comprises of the community, its habitat and the natural ecosystem. The community refers to the group of people who are benefiting from the biorights intervention, whereas habitat refers to the human habitat wherein the community performs their livelihood or other income generation activities. Both the community and the habitat are an integral part of the broader ecosystem around them. If we look into the integration of these systems (Fig. 3.1), we find there are a number of circularities amongst the components and the activities develop an intricate matrix of relationships. This helps to accommodate multiple aspects and features those are drivers and determinants of the system in entirety. We will discuss a few. Primary productivity in one way generates livelihood and as well provides food security, whereas returns from all income generation processes are invested in building societal securities, risk mitigation, acquiring basic amenities and other developmental activities. On the other hand, the community, as a component of biorights intervention, uses the resources and services of the habitat that they live in for both farm and non-farm sector activities. Their responsible usage keeps the habitat health and therefore sustains the ecosystem to intensify its resources and services. Communities engaged in biorights intervention also repair and restore the habitat using ecosystem-based adaptations and natural solutions. Better habitat health and enriched soil functions assure sustainable production and food security, while more intensified ecosystem services, e.g., flood control, ensure better risk mitigation and community resilience. As the community restore and reuse the habitat, gradually they acquire their natural rights on it, but in actualizing restoration and responsible usage they loose opportunities of resource accrual and some other development facilities thereto. For example, to prevent overgrazing the communities may have to size down their herds or forgo the opportunities of developments like a primary school for children or a veterinary hospital in the pastures that may lead to some urbanized encroachments further diminishing their livestock size. This opportunity cost needs to be compensated and biorights compensates this with Payments from Environmental Services (PES). A highland ecotourism or cash crop farming around the pasture can be some simple examples to transform the PES to a fiscal return for compensation. All such additional interventions are adaptive in context to the ecosystem and has to be a natural place-based solution. Therefore, this compensation gets endorsed by the biorights program as a community right. This is furthered with strengthening of local institutions, financial inclusion and local economic growth. The interesting aspect of this model is its self-regulatory mechanism. An in-built feedback inhibition control always tries to attain a dynamic equilibrium. One such point of control is that irresponsible usage of habitat resources and services gives negative feedback as evidenced by diminishing ecosystem services and compensation of opportunity costs will be withdrawn inevitably. For example, intensive fisheries damage wetland habitat and may increase the nutrient load leading to eutrophication, depletion of dissolve oxygen and increase of biological oxygen demand. This may impact on the aquatic biodiversity, ecotourism potential of the wetlands and as well may coerce overfishing in river capture fisheries owing to the subsequent fall in fish productivity in the wetlands. On the contrary, extensive ecotourism may also cause fall in productivity leading to almost similar effects. Wise users need to know these trade-offs and synergies to act in a synchronized way for sustainable production. Such tipping points also exists between intensification of ecosystem services and payments from ecosystem services or investments in developmental activities and compensation of opportunity costs.

Thus, the design thinking approach in biorights model provides a resilient yet flexible paradigm to cope with the changes in the local surroundings and randomness of activities within the paraphernalia, which is crucial for sustainability in the climate milieu. We will discuss few case studies pertaining to this from South and South East Asian nations in the following chapters.

3.4 Biorights – Discussing Few Case Examples and Global Forms

3.4.1 A Case from Central Luzon, Philippine

In Luzon Philippines, groups of student volunteers from State Agricultural University Central Luzon have teamed up with local fisher's communities to conserve patches of mangroves in nearby coastal areas through community forestation and conservation of habitat. It sustained a small nursery of two species of mangroves namely, *Rhizophora* and *Avicennia* in a government owned piece of coastal land. This small endeavor has grown into a robust young forest due to the wise management of the team that includes a leader in each site, locally known as the "owner" or manager. The manger's residence is located in front of the reforested mangrove while that of the remaining community volunteers, mostly females above the age of 45 years, are nearby. Community livelihood herein has been very much associated with the sea and they were socio-ecologically included in the category of fisher folks. As part of their caring and maintenance strategy, they fenced the mangrove forest with bamboo poles except the side facing the open ocean. Elevated bamboo bridges were made to interconnect various sections throughout the mangrove. Bamboo stairs located in particular nodes of the bridges facilitate in going down to particular stations. A thatched community hall supported by towering stilts was constructed to provide shelter to the volunteers or visitors in case they want to stay there as observers or use it as a meeting venue. The proper care rendered by the team brought favorable maintenance to the mangrove forest that included getting rid of dead leaves and stems off the mangrove trees and picking them up from the ground while covering areas believed to be hatchery spots of edible species with dead coconut leaves and stalks. They also carefully trod the mangrove grounds so as not to step or disturb the developing eggs and larvae and get rid of stray animals as well as garbage that were not properly disposed especially in areas near the road. This sustainably intensified the ecosystem services of the area and created new services, as well.

The edible organisms that thrive successfully in this mangrove forests are species of shelled molluscs and the crab, *Scylla serrata*. The ecological parameters such as prevailing ranges of temperature, salinity and all other physical and biological components seem to suit the growth and development of the edible crab, *Scylla serrata* and mollusks leading to their continuous production and harvest. These served as food for the community volunteers especially during natural emergency. Later this was sold to the outsiders which eventually brought income to the team. Crabs are available for sale throughout the year. This may be supported by yearlong spawning of high number of females (Ali et al., 2019). The teams gradually established a netted pond and allowed commercial brackish water fishes to grow and proliferate. In exchange for maintaining the mangrove, the communities were given crabs and molluscs by the team along with fishes those were sometimes given for free because of their abundance in the area. The natural ownership right emerged among the communities and the local municipal administration partnered with them to buy back this coastal conservation efforts and brought in incentives for the communities in kind.

The partnership expected that the sites could be very good component for ecotourism purposes. This will ultimately add additional income to the communities and the volunteers though there were no iota of doubt that the income is directly linked with the habitat health and therefore sustaining the conservation efforts are crucial. Introduction of salt tolerant rice in near vicinities of the village through capacity building from the agricultural university became another economic opportunity in mangrove areas with extensive peripheral grounds. They were earning money and at the same time helping in the maintenance of the ecosystem of the reforested mangrove making it ecologically healthy and economically sustainable. The municipal government and the university are behind the entrepreneurial initiative through conservation targets and mandates, as well as the proper maintenance of the habitat. This mangrove area is now not only a source of food security for the community but a pioneering replicable model for economic sustainable development and healthy ecosystem as well.

The biorights component in this case is originating from the transformation of ownership to stewardship. The community and the team are in a position now that the cost of the opportunities that they forgo for conserving the mangrove patch can now be recompensed through alternative earnings and in-kind support from the partners, like capacity building, as well convergence of resources through municipal Schemes.

3.4.2 A Case Example from Nigeria[1]

The largest wetland in Africa is the Niger Delta in Nigeria that has three sites listed as Ramsar Wetlands of international importance, whereas it is also the third largest mangrove forest in the world with enormous conservation significance. The region is rich in biodiversity as well as its oil and gas resources. Wetland ecosystems play a critical role in supporting the livelihoods of millions of people inhabiting the delta. The region is home to nearly 30 million people, and 60% depend directly on the services provided by the environment – such as fish and clean drinking water. The anthropogenic pressure, however, is degrading the ecosystem owing to unsustainable practices and continued pollution from oil spills. The intervention therefore, prioritized the conservation and restoration of wetlands for achieving both livelihood sustainability and biodiversity protection. Nigeria being the eighth largest oil exporter in the world, it is imperative to say that development of the abundant oil and gas resources has created a host of environmental and social challenges. Wetland's degradation is a serious ongoing problem, caused by pollution from oil spills and gas flaring, the population influx attracted by the oil and gas industry, poverty, uncontrolled exploitation of the forest (wood), over-fishing, and various poorly planned infrastructure developments.

Historically, little of the oil revenue generated in the delta has made it back to be reinvested in local communities. The standard of living for Niger Delta communities is very low and basic requirements, such as access to clean water, health facilities and education, are lacking for most. In order to address some of these issues, there is an urgent need for development initiatives to integrate the livelihoods of local communities with ecosystem restoration. Despite the significance of the Niger Delta as a wetland, environmental improvement programmes by the oil and gas sector have been lacking for a long time. A major shortcoming has been the insufficient understanding of the value of wetlands and their importance to different stakeholders. Without an awareness of the links between wetlands and livelihoods, development initiatives that aim to tackle the challenges faced by communities may not be sustainable or successful. The biorights intervention was implemented through a local partnership initiated and mobilized by Wetland International and Shell that worked to:

- Demonstrate how the restoration of wetlands can be linked to improved livelihood opportunities in local communities. To achieve this, we studied the wetland types and mapped the ecosystem services at different locations in the delta to analyse their vulnerability to different activities;
- Engage local communities to integrate wetland values into development plans, with ambitions to enhance ecosystems and sustain livelihoods through incentive systems such as Bio-rights;

[1]This case study has been adopted from Eijk, P. van & R. Kumar, 2009. Bio-rights in theory andpractice. A financing mechanism for linking poverty alleviation and environmental conservation. Wetlands International, Wageningen, The Netherlands.Published by Wetlands International.

- Strengthen the capacity of local civil society partners to implement ecosystem-based restoration and expand the scope of the most successful projects;
- Integrate wetland values into the government's sustainable development planning;
- Change the practices of the wider oil and gas sector in order to understand wetland values and protect, sustain and restore them.

While this work happened under this partnership, partners were not directly involved in the clean-up of oil pollution, but included the oil industry as an important stakeholder for their new ways of planning development and improving the condition of wetlands and water resources to benefit both biodiversity and human well-being. The *Abobiri, Obia-yagha and Opume* communities in the Niger Delta gradually changed the way they manage their wetland environment. Fifty-eight community member-groups, provided with access to micro-credits, stopped wetland-detrimental livelihood practices like mangrove deforestation and switched to wetland-friendly livelihood practices such as fish, periwinkle, plantation and poultry farming. In addition, they contributed to wetland restoration through wetland tree nurseries and restoration, tree planting and clearing of waterways overgrown by invasive plant species.

3.4.3 Case Examples from Bhutan Himalayas

3.4.3.1 Community Forestry for Sustaining Life and Environment

The Himalayan kingdom of Bhutan in eastern Himalayas extends over 46500square Km with elevations of 100 m to 7550 m. The annual rainfall ranges between 500 mm to 5000 mm across the altitudes (WWF, 1999; Dey & Sadruddin, 2000). It is situated in the confluence of three major biologically endowed zones viz. the palaearctic realm in Asia to the north of Himalayas and further west, the boreal zone along the Himalayan range and the palaeotropic distribution found in the Indian subcontinent and South East Asia. This confluence positions Bhutan as one of the 'hottest hot spots' of biodiversity in the world (Lobzang, 1998; Norbu, 1994; Grierson & Long, 1983) through the admixture of bio resources in these zones. With an intact forests stand of 72.5% of the total land area and a growing population (3.1% per annum) the Royal Government of Bhutan has designed new strategies for sustainable development. These initiatives highlight reciprocity, equity and participation to ensure economic growth, without compromising on environmental conservation (8th Plan Doc.; Dey, 2002).

In this fragile mountain environment farming systems play crucial roles in providing for the needs of the local population. Sustainability in these contexts calls for proper management of land, water and forest resources. The concept of social forestry was accordingly introduced in Bhutan in 1979, by a Royal Decree. This included various forms of community forestry that emphasized traditional systems of resources utilization and indigenous knowledge for sustainable management of forests (Messerschimdt et al., 1996). In 1993, decentralization of the social forestry extension programme to the district level and the start of the Third Forestry Development Project (TFDP) accorded social forestry high priority and enabled local forest management activities on government land. This was possible through the participation of local communities deciding and implementing in conservation and optimal utilization practices.

Bhutan is endowed with more than 902,000 hectare of operable forest, containing an estimated standing volume of 184 million cu. Mt.; more than 10 cm diameter at breast height. The annual allowable cut is only 0.64%, whereas afforestation and regeneration rates ranges between 1.8% and 3.3%, PIS estimated annual increment in hardwood of 3.7 to 5.96 cubic meters in broad leaf forests and 6 to 6.88 cu. Mts in conifers. Information on the kind and quantity of forests is presented below. (Table 3.1).

The NTFPs in particular in the forest resource of Bhutan (known as **Dzong GyelKhab** – the den of medicinal plants in the past), is reportedly adequate for the rural communities. (Dorji, 1995; Namgyel, 1996). Important medicinal plants include *Cordyceps sinensis* (a fungus growing over *Leptropteris* butterfly larva), *Panax, Taxus baccata, Belledona, Aconitum, Swertia, Herachlium, Zanthoxylum* etc. About 141 species of orchids are reported from Bhutan Himalayas. At least 25 of these are in the IUCN red list. Forests of Bhutan also host many such endemic plant species as the Blue Poppy (*Meconopsis grandis) Primula sp. Ceropegiadorjei*, and animals as the Golden languor (*Panthera uncia*), Black necked crane (*Grus nigricollis*). Many Gondwanan flora and flora of African origin (*Lobelia nubigiana*) can also be spotted here. Out of the 52 species of Rhododendron, *R. kesangii* is also endemic to Bhutan Himalayas (Sadruddin1997; Lobzang, 1998).

Bhutan has a predominantly biomass-dependent economy with agriculture being the main subsistence occupation of 87% of Bhutanese people. The total arable land

Table 3.1 Showing %age of different forest covers

S. NO.	Forest types	Dominant speceis	Area (%)
1.	**Conifer Forest**	*Pinus bhutanica, Larix griffithii, Abies densa, Picea spinulosa, Tsuga dumusa, birch and Sub-alpine bamboos.*	**26.5**
(i)	Fir		8.6
(ii)	Mixed conifer		12.1
(iii)	Blue pine		3.2
(iv)	Chir pine		2.5
2.	**Broad leaf Forest**	*Quercus sp. Alnus sp. Shorearobusta, Duabanga grandiflora, Melia sp. Aitlanthus grandis, Bombax ceiba Magnolia cambelii, Micheliasp etc,*	**37.7**
(i)	Broad leaf + conifers		34.3
(ii)	Conifers		03.4
3.	**Forest plantation**		**0.2**
(i)	Conifer plantation		0.1
(ii)	Broad leaf plantation		0.1
4.	**Scrub Forest**	Dwarf juniper and *Rhododendron sp,*	**8.1**
Total			**72.5**

in Bhutan is approximately 3146 Km^2, which accounts for 7.7% land of the entire kingdom. 31.1% of this land is under dry land cultivation and 38.1% is under shifting cultivation (Tseri). Wetland and irrigated cultivable land are only 12.3% of the total arable land and the remaining of it has mixed cultivation practices (Gyamtsho, 1998).

The study suggests that 87% of the area under rice cultivation is through local or indigenous varieties. The rural inhabitants collect about 140 species of plants of diverse ethno-botanical importance from the forests for their livelihood (Dey & Pankaj, 2003). The National Institute of Traditional Medicines alone use more than 300 plant species. Handmade paper industry, lemon grass oil extraction units, resin tapping, turpentine oil and cane products are some of the most common small-scale industries of rural Bhutan.

3.4.3.2 Status of Community Forestry in Bhutan

The crux of Buddhist belief in Bhutan is *'tendrelgichoe'*, the interdependence among all forms of life and dependence of life on forest resources. This emerges from the conservation tenet of Buddhism. Forests and forest products are venerated in daily life, traditional practices, and social rituals in Bhutan. This spirit is reflected in community forestry that signifies common values of the society, ecology, economy, morality, politics, spirituality, integrity and right thinking (Wasi, 1997). It is essentially a locally relevant short-term enterprise for long term sustenance but a broad based land use system encouraging high degree of participation and involvement of local people (Messerschimdt et al., 2000). The Forest act of Bhutan 1995, Social Forestry Rules 2000 and Forest and Nature Conservation rules of Bhutan 2000, define community forest as "any area of government reserve forest designated for management by a local community in accordance with the provisions under section 36 of these rules." Community forestry in Bhutan refers to the control and sustainable management of local forest resources by the users (Gilmour & Fisher, 1996). The principal objectives of the community forestry in Bhutan aligned with the Buddhist philosophy aims to –

(a) Promote active involvement of local people at all cross-sections of the society for forest resource management, sustainable development and equitable sharing of benefits in order to improve rural economy and living standards.
(b) Enhance efficient utilization and protection of forest biomass and other forest resources, maintain and improve biodiversity and ecological functions of forest lands.
(c) Build on indigenous knowledge and traditional practices of natural resource management as well as strengthen the institutional and technical capacity of forest management groups.
(d) Utilize an extension dialogue for two way learning and teaching for establishing partnership and cooperation among developers and users for promoting social equity.

3.4.3.3 Policy and Practice

Social forestry rules 2000 defined a community forest as a suitable forest area located within reasonable distances from the local settlements. These are not more than 2.5 hectares per households comprising degraded and good forest in equal proportion. Community forestry in Bhutan has two forms. The first form is about reforestation and afforestation on degraded, degrading and barren lands. The second form is on a mix of degraded and natural forests, designed by the district extension officers, a local management group and a local executive committee with reference to an agreement or plan.

The 2nd June of every year is religiously observed as the Social Forestry Day in Bhutan. The religious flag-poles that are a ubiquitous symbol of Bhutan's Buddhist heritage; made of only two conifers, *Pinus bhutanica*and*Cupressuscashmeriana*, endemic to Bhutan forests. 'Ridam', is one common forest-oriented custom that restricts access to specified forests during certain seasons of the year. 'Ladam' is the ritual closure of biological corridors and mountain pass across forest areas. The explicit rules of ridam and ladam are defined by the local community and people. "Kidu" or "Solera" are one-time royal grants or gifts to individuals, institutes, organizations or monastic bodies that allows the recipient to utilize the forest resources. Two fundamental conditions dictate the success of participatory community forestry in the locality. They are the clearly stated interest and willingness of local villagers and trust and confidence in local management by the government. Both are based on the notion of 'civil society'- a fundamental component of the local community that ensures mutual support, social services and any other form of participatory citizen action (Fukuyama, 1998). Two important case studies highlight here, how community forestry in Bhutan progressed under the Forestry Development Programme 2008. However, there have been some predicaments in practice and policy in the dragon kingdom, which we will discuss after these.

(i) *The Dozam Community Forest* in Dramtse Block of Monger district, in Eastern Bhutan had got heavily depleted due to timber and fuel wood harvesting, unplanned extraction of lemon grass oil, uncontrolled forest fire and overgrazing, leading to regeneration problems of standing vegetation. Distorted vegetation dynamics and deforestation had exposed the forest floor surface to wind and water erosion. About 63,380 seedlings were sown through community partnership program and 56% survival rate was achieved. Significant outcomes included broadcasting of 1.3 Kg Chirpine seeds, restored regeneration in oak and *Castanopsis*, prevention of forest fire by the endeavours of local community, licensing of the lemon grass oil production units etc. In compensating the costs of man-days contributed by the communities, part of revenue earned by commercial traders from pine oil extraction and license fees obtained from lemon grass oil production units were contributed by the local administration into community forest management fund, by-laws were also made to control grazing and over exploitation of resources. Community forest management group assessed progress and incorporated the suggestions and advices of the users, as well. The sustainability of both trade and ecosystem were assured through a mutually beneficial arrangement.

(ii) In *Chaling* Village of Trashigang district of eastern Bhutan, the forest was getting depleted rapidly alongside of dropping agriculture productivity. Scarcity of drinking water and acute shortage of fodder in winter were the key problems. The site was inaccessible from the highway and lacks in fuel wood. It was possible to overcome by training few users in the technicalities of resource management. They could be trained at the nearest Renewable Natural Resource Research Centre, in Khaling, eastern Bhutan. The supply of good seeds and seedlings could be expedited and lopping of natural fodder trees were minimized. The local administration could provide drinking water and fuel wood and construct feeder roads for better access. There has been significant progress in community forestry in Chaling since then. These could be strengthened through some policy support for legal procedures to develop community forestry in woodlot and pastures. A strategy to restrict agro-pastoral activities in surfaces with slope gradient 45° and below, that was recommended in the 9th Plan Document 2002 was also reviewed and 'slash and burn' type of crude cultivation practices was regulated.

3.4.3.4 Issues for Further Planning

Conservation and natural resource management must go hand in hand for better people-environment relationship. The Royal Government of Bhutan has to tread carefully with its policy instruments to ensure this. Bhutan has the advantage of starting late but also has the handicap of lack of experience and paucity of technical knowledge. Dependence of rural people in Himalayan belts on natural resources demands a more systematic planning if a large proportion of today's biodiversity is to exist in future for increasing number of people and their demands on natural resources (CIBC report 2000).There is least doubt that informal border trade is significant for Bhutan. This has resulted into a concentration of population around the borders. As a natural outcome to such concentration, there is a reduction in arable land vis-à-vis the growing demands of agricultural products in the border areas. This has led to an excessive use of chemical fertilizers and pesticides. On the other hand, having no standards set to monitor the transaction of 'environmentally sensitive goods', Bhutan has to hoard wastes and packaging materials coming from India, Bangladesh, Tibet and Nepal that are not eco-friendly. Consequently, the foothill ecology is at a stake and soon there may be issues pertaining to land-rights and forest rights. Bhutan, the land of thunder dragon, has to explore the possibility of identifying niche markets within the sub-region and also outside the region, so that the trade avenues may be expanded. The potentiality of such an expansion will be more if Bhutan gives a consideration to the emerging environmental standards. In the climate milieu, it is in the interest of the Himalayan kingdom to develop its own standards based on sustainable production of commodities. Such a linking of trade with environment will be in conformity with Principle 12 of Rio Declaration.

This is likely to bring a double advantage of trade diversification and environmental protection. The potentiality can be explored in the areas of organic agricultural products (OAPs), traditional handicrafts, manufacturing, beverages, herbs-based pharmaceuticals and products based on bioprospecting etc. In that neo-economic uprise, the environmental services will be better restored for the people and shall be able to compensate their lost opportunities therein.

Along with environmental standards in trade, it would be in the interest of Bhutan to device a machinery for 'carbon advantage' studies by estimating the sequestration potentials of its conserved forests that can help in setting its own carbon offset marketing strategies in leveraging the global carbon financing mechanisms. While addressing the current situation, Bhutan has to look up for more feasible options to sustain development, environment as well as happiness of the people. Biorights is a befitting mechanism in this traditional socioeconomy of indigenous people. Integration of efforts at the cross-sectorial level within the eco-region is one option whence problems related to environment can be better reviewed and sorted regionally than importing solutions or hiring technology from the other globe. The only advantage of starting late can be learning and sharing the experiences. This is the area where the scope of biorights can be explored by the international development agencies for Bhutan.

3.4.4 Ecotourism in South Asia: Combining Wetland Conservation and Business

The simple Indian mythological stories, lullaby or common ballads pertaining to worship of nature, appear to have metamorphosed into socio-economic and environmental jargons through research, seminars, workshops aided by high technology. Otherwise, there is hardly anything new in such concepts as wetlands, ecotourism, biodiversity, that we are used to today. They are all correlated on the basic premise that, biological diversity of the wetlands can be protected through sustained use of tourism. But what is new today, is the awareness, that the hopes and aspirations of the future generation can never be sacrificed for the sake of fulfilling the ambitions of those who have undermined the contributions of nature, indulged in unscientific planning and economic development.

Wetland and its Biodiversity:Wetlands are among the richest ecosystems on this planet, only next to the tropical rain forests. The World Conservation Monitoring Centre has estimated about 5.7 million sqkm of wetland, roughly accounting for 6% of the earth's land surface. The greatest proportion is made up of bogs (30%); fens (26%); swamps (20%) and floodplains (15%) with lakes accounting for just 2% of the total. Mangroves with some 80 species of trees and shrubs cover about 240,000sqkm of coastal area and an estimated 600,000 sq. km of coral reefs remain worldwide (Ramsar Secretariat, 2010).

Wetlands including (inter alia) rivers, lakes, marshes, estuaries, lagoons, mangroves, seagrass beds and peatlands are among the most precious natural resources on earth. Wetlands occur in every country from the tundra to the tropics. Currently 39 World Heritage Natural Sites contain major wetland values and an additional 38 sites have secondary wetland values. The 77 total sites in this inventory represent 50 countries and range from a 19 ha nature reserve in the Seychelles Islands to a 9million ha boreal lake in the Russian Siberia to the 35million ha Great Barrier Reef in Australia. Less number of World Heritage Sites with significant wetland values currently exist in Central Asia, the Middle East, the Polar regions and the South Pacific (IUCN, 1997).

Unlike the coastal wetlands, Himalayan wetlands bear unique characteristics in South Asia, especially with the inclusion of countries like Nepal, Bhutan, and Sikkim. Nineteen major rivers and innumerable wetlands originate from the Himalaya – of these the Indus, the Ganga and the Brahmaputra are the largest. The predominant wetland types in the eastern Himalayan region are riverine marshes or beels, freshwater swamps and lakes. By virtue of their productivity, floodplain wetlands including beels, jheels, pats, constitute important habitats for fish and fisheries. While fisheries constitute an important economic activity, the Himalayan wetlands provide for food, fodder, fuel and several other products. The reeds and cattails are used for thatching and some plant species are harvested for medicinal uses. The wetlands in the eastern Himalaya exhibit greater diversity with regard to flora and fauna. Freshwater swamps along marshes support a large number of threatened species and are considered to be among the most important wildlife habitats internationally. Some of the important wetlands located in the region are: Loktak Lake, Manas National Park, Laokhawa, Orang and Sonai Rupai sanctuaries, Kaziranga National Park, Lili Sanctuary, Mahao (Mehao) Sanctuary and Namdapha National Park. Loktak Lake is the largest wetland in the eastern Himalaya and plays an important role in the ecological and economic life of Manipur.

In addition to the exceptional biodiversity, wetlands are also of immense social and economic value for both traditional and contemporary societies. In terms of producing social values, they provide clean water to towns, protect the shore lines from storm, sustain fisheries, invertebrates, amphibians, reptiles and mammals as well as millions of both migratory and sedentary water birds. Once proclaimed as waste lands fit for drainage, wetlands are now recognized among the most productive areas and potential values have been associated with them. From an economic point of view, they act as important tourism destinations as visit to these areas provides the aesthetics of pleasure and satisfaction to the tourists.

Several communities depend on the wetlands for drinking water, food, fish and other products for their sustenance, either directly or indirectly. . Unlike the forest ecosystem, most wetlands require a certain amount of human intervention which, if, used judiciously, will establish a symbiotic relationship between the wetland and human community. Hence the challenge to the society is to develop relevant solutions towards this goal. The Convention on Wetlands of International Importance, especially as Waterfowl Habitat, held at Ramsar in Iran in 1971 deserves mention in this regard. This international treaty, also named as 'Ramsar Convention' assists the countries to manage their wetland resources, in order to get the environmental benefits and also improve the living standards of the communities.

Given the low level of environmental awareness and irresponsible attitude in general, mass tourism becomes accountable for loss of wild life, habitat, deforestation and pollution in the regions of travel and tourism. Alternative form of tourism such as ecotourism seeks to avoid such outcomes by pursuing tourism in a sustainable way and depending more on ecosystem's well-kept recreational services. Herein, ecotourism comes in proximity of biorights and it can be easily used to compensate the opportunities lost by the communities in keeping the good health of the habitat. It follows a middle path between tourism promotion and environmental conservation with a trade-off calling for careful choice of interventions to ensure economic and environmental gains. Importantly ecotourism has achieved institutional legitimacy. The International Ecotourism Society (TIES) defines ecotourism as "responsible travel to natural areas that conserves the environment and improves the welfare of local people". It includes such aspects as sustainable use of resources and minimize adverse impacts on environmental, ecological, cultural and social domains in order to maximize benefits for the local community. This minimizes damage to the biological and cultural aspects of the environment and promote a better understanding of the natural and cultural history of the region.

Since long, ecotourism has been growing in the range of 20% to 34% per year, as one of the most privileged business houses (TIES 2005). South Asia has been the melting pot of diverse cultures and though it has immense potentials of ecotourism, it lacks carefully made, well-crafted sustainable development strategies in the anthropocene, where pressure of population and pressure on species are both elevating fast. Hence community collective in systematic monitoring of the critical areas with cooperative approach to cope with the problems is the key concern of this region. To ensure environmental sustainability it is necessary to integrate the principles of sustainable development in official policy. South Asia is a region with over 1.3 billion people, the largest concentration of humanity in a single geographic region that amounts to 23% of global population, where the total wetland area of 96970sqkm is estimated to be of an economic value of $1.8 billion per year (SAEFL, 2004). The developing economies of the South Asian region can leverage economic growth by sharing nature services of wetlands, common tropical climate, proximity and most important, the cultural linkages of the region for marketing their ecotourism products that can help alleviate regional poverty. Policy planning, however, has reversed the picture altogether.

In Sunderbans, the largest delta in the world and a coastal Ramsar site spanning both Bangladesh and India, the fishermen and honey collectors were often attacked by the tigers during their forays to the mangrove forests. This conflict is owing to the locking of their access to the natural resources in the pretext of conservation by declaring it a 'Tiger Reserve'. The natural rights of the fringe population are being denied leading to irresponsible resource usage by the ultra-poor, risking their lives amidst man-animal conflict for the sake of livelihood. The relation with the local people could have been improved considerably by the formation of Forest Protection Committees and Eco-development Committees (Vyas, 2005). This has been still better in Nepal's large trans-Himalayan wetlands as the Rara, Phoksundo, Gokyo and Tilichu, that has great tourist value being the migration fronts and staging areas of many water birds taking the Central Asian flyway (Kirby, 2010). The distribution of indicator wildlife in scenic wetlands of Nepal has helped these wetlands to play a significant role in ecotourism, contributing to local economy.

Likewise, in Maldives since the coral reefs and three-quarters of world's reef fish species are facing the threat of extinction due to dredging, bleaching, increase in cyclone intensities and other man-made methods, the Maldivian Ministry of Tourism through a private-public partnership policy encouraged 'Four Seasons Resort' management at Kuda Hurra to launch the reef-ball project to rejuvenate damaged reefs around the resorts with the use of reef-ball structures designed to facilitate coral growth. The first of its kind in Maldives, the use of reef-balls have increased the growth of corals tenfold and coastal ecotourism subsequently. This is a win-win example of convergence, innovation and partnership that inspires the biorights interventions as well.

Local people's participation in ecotourism in wetland management is of prime importance since they are an integral and a dynamic component of the ecosystem. In this region, the natural resources at the subsistence level are manipulated and controlled by the local people, even though, the land tenure is determined by the government. Since the region's wetlands contribute to the livelihood of a high %age of the local community through floodplain agriculture, open water capture fisheries, swamp forest plant products, domestic water supply, and provides a means of transportation and communication, people, in and around these wetlands, have evolved indigenous management systems. For example in Bangladesh, the term '*Garubala*' describes the system of community management of livestock. And '*Nitimala*' is the people's participatory approach to natural resource management. Hosting 173 bird species both under common and uncommon categories, the Uchalli Wetlands Complex in Punjab, Pakistan, has also involved common people for the wetland management.

The specific areas in which there is an urgent requirement for public participation and needs to be enhanced in the South Asian context are (a) Lowland floodplain and forest management, (b) Sustainable utilization and protection of wetland weeds and wildlife and (c) Integrated management of wetland ecosystems. Ecotourism will promote linking people with nature, make users understand and conserve wetlands resource bases for tourism and add additional conservation values to the wetlands.At a point of time, in the developing economy of the region when traditional conservation through enforced protection of natural resources are being questioned for its effectiveness and social impacts, strategies like ecotourism offer considerable potential for integrating conservation with development. Because in these countries, particularly, governments lack sufficient funds to manage and protect natural resources, that are under substantial pressure due to hunting, farming, logging and fishing.

Ecotourism as a tool for conservation also means economic sustainability, support for conservation and local people through the revenues of tourism. Ecotourism projects may either create sustainability for the wetlands, local tour operators and

MSMEs as well, or the projects may fail if such ventures are devoid of keen sense and assessment of the economic viability of the project. Several socioeconomic issues, like eviction of indigenous people to promote ecotourism destinations, the 'fees and fines' policies to protect and preserve biodiversity, the identification of genuine ecotourists and nature tourism operators have raised questions on the impact of such vague rules and definitions on the ecological character of the wetlands and can also affect the livelihood of the local community who are dependent on the resources of the wetlands, creating a conflict interface with the management. Alike conservation strategy, Ecotourism, in the wetlands also corresponds to the current paradigm of 'no consumptive use' of wetland resources.

3.4.5 Economy and Ecology of Pastoral Nomads of Eastern Himalayas

Brokpa tribes are primitive agro-pastoral highlanders of eastern Himalayas living in the Merak and Sakteng gewogs (blocks) of Trashigang Dzongkhag (districts) in Eastern Bhutan, occupying eleven villages in Sakteng gewog with 327 households and 2119 heads as well in four villages in Merak gewog with 213 households and 1918 heads. They shares border with Arunachal Pradesh of India in the north and in the east and practice seasonal transhumance in this pristine ecology of eastern Himalayas.Collectivism and cooperative living are the sole character of the Brokpa society and they maintain a homogeneous socio- cultural category. Animal rearing has all along been the primary occupation of the Brokpas. Weaving woollen goods and bartering milk products with lower valley inhabitants are a part of their livestock economy. Herding and trading together account for 66.17% of the total work force. About 67.30%Brokpa families practice transhumance with a maximum of 91% families in Merak. *Drukor* – a barter trade link between Brokpas and their Sharchopnepos (village hosts in Eastern Bhutan) is their integral economic practice followed for centuries. Products such as cheese, butter, homespun blankets and floor mats are bartered by Brokpas for rice, barley, chilies and other edible items. Agriculture adds to their normal husbandry activities. A summer crop of buckwheat and mustard in the upper valleys, and maize, wheat and barley in the lower valleys are cultivated in Brokpa villages.

Brokpas are basically yak herders. The extreme altitude and harsh climate make it hard to grow agriculture crops. Therefore, their rigorous lives and their very survival depend on nature's services in entirety for these animals. The abode of Brokpas stretches from 1423 to 4516 m in the Greater Himalayan ranges. The flatter and fertile alluvial fans are used by the Brokpas for buckwheat cultivation in the floodplains of river Saktengchhhu that originates from Zangpin Sumglaciallakeat 3980 m near the Arunachal border, while river Bangpachhu, the second important tributary of river Gamrichhu takes origin from Dangzongglacier at 4465 m. Merak is an upland valley more or less like a plateau approximately 4 km long and 1 km wide and is endowed with alpine vegetation and pastures. The mixed broadleaved forest of lower valleys merges into forests of oaks; rhododendron, birch, fir, juniper, spruce, and mountain cane. The timberline is found at 3980 m. Beyond this lies the vast expanse of the alpine meadows. The landscape provides places of raw natural beauty of high alpine grasslands, rolling hills and towering crags with only small patches of habitation with minimal deforestation. The pristine mountain environment has not succumbed to developmental pressures of modernization. In brief, the land and people here had existed in harmony over the centuries.

Brokpas are mainly pastoralists belonging to the same band of highland nomads that are strewn across Central Asia. The predominantly mountainous environment and a characteristic cold climate afford but very few opportunities for the cultivation of crops or exploitation of other resources for economic betterment of the community. The abundance of high-altitude alpine pastures for summer grazing, forest meadows and bamboo grasslands in lower altitudes for autumn and winter grazing gives rise to full -scale pastoral activity. The wealth of these pastoralists originates from the yak herds. The Brokpa economy revolves around the yaks, and trying to separate the Brokpas from their yaks would tantamount to putting their lives and destiny in great peril. Merak has completely a pastoral economy whereas partial dependence on summer crops like buckwheat and mustard supplement the livestock economy in Sakteng. In the lower Brokpa regions, maize, wheat, barley and mustard are cultivated. Many Brokpa families in Sakteng own farmlands in the lower regions.

Brokpa community considers those who do not own cattle as poor. The poor Brokpas work as carpenters, blacksmiths and farm laborers or as yak herders for the rich families. Blacksmiths are mainly Tibetan migrants settled in Sakteng. Many such blacksmith families have also settled in Radhi and Phongmey villages. Shifting their camps all the year round in many villages of Trashigang Dzongkhag, they make copper ladles, spoons, kettles, pots and pans, and in turn gather rice, maize etc. from the villagers. In general, Brokpas have practiced transhumance and established trade contact with lower regions of eastern Bhutan and Tawang region of Arunachal Pradesh. During winter and spring they barter their yak products mainly cheese and butter with the food grains. Today with increasing exposure to the markets, the barter trade is slowly getting replaced by the cash-based economy. The following table (Table 3.2) shows the pastures under the possession of the Brokpas in Merak and Sakteng in 2009.

Brokpas have devised their own ways of maintaining animal health. In winter, starting from the 25th day of the tenth month of the Bhutanese calendar, the people of Merak and Sakteng do not milk their yields. After 12 days of birth, the calf is

Table 3.2 Pasture coverage in acres for Merak and Sakteng 2000

S. No.	Pastures	Merak Gewog	Sakteng Gewog
1	Summer pastures	15203.63 acre	20783.46 acre
2	Autumn pastures	3220.21 acre	6562.76 acre
3	Winter pastures	5346.10 acre	79114.80 acre

separated from its mother and is set free all day after milking in the morning (RNR Newsletter, 1999,p.9). To support livestock activity in the Brokpa region, the Royal Government of Bhutan has introduced breed improvement, sterilization of scrub bull to reduce population of unproductive cattle and health service delivery programs. These supports have incentivized their nature-based lifestyle.

Yak meat is a luxury food for many and people generally do not kill productive yaks. It is only old yaks and sheep that are slaughtered for meat and skin. Yak hair and sheep wool provide important raw materials for Brokpa clothes, caps, blankets, bags, ropes and their dress. Regrettably woollen manufacturing has still not acquired a central place in the Brokpa economy. Clothes from wool are woven mainly for their own domestic use. In Saktenggewog, more than 30 pastures had been identified and mapped for their LUP studies. Four major categories of land use were dominated by forests, which share 59% of the total recorded land area. Pasturelands come next to the forest having 36.9% share in the total land recorded area. Most of the pastures have been registered (34.2%) with the local administration and only 2.7% is described as natural or non-registered. Agriculture occupies a very insignificant proportion (2%) of the land use feature. Area under settlement is 4.4 ha. i.e. only 0.001% of the total area. Other categories include 2.1% of the total land with a major share of 1.6% under rocks and crops. Besides the open pasture areas in the alpine zones, the forested areas especially the oaks, rhododendrons, sub-alpine fir and juniper forests provide important grazing lands for high altitude livestock in this part of the Brokparegion. Pastures here are mainly forest meadows. The main grazing lands of the Brokpas are, of course, the vast expanse of the alpine meadows.

Pastures are the main domains of the Brokpas. Three main areas as dictated by seasonal control make up their domains: summer, autumn and winter pastures. Brokpas have, therefore, a defined seasonal cycle of movement within these three pasture regimes located between 2500 to 4500 m and categorized by a harsh climate with snow in winter and abundant rain in summer. Summer pastures throng the upper reaches of the region and Brokpas move to these pastures in April and stay till the end of June, that is, as long as the green pastures last. The herders drop lower and lower in pursuit of the fast 'disappearing grasses and reach the autumn pastures which are normally located on the mid mountain slopes and their stay here last till the end of August. As cold weather sets in the upper reaches, they slowly move towards the winter pastures that they use from September through December. Winter pastures are located on snow free areas where grasses are available in cold season. May and October constitute the peak months of their summer and winter migration when most of the households set out to move. The yak herders live in stone houses and in thatched huts in winter and in summer they camp in the pastures.

Brokpas have their registered pasturelands locally known as *'tsamdrog'*.Some pastures are officially leased out to them for free near the villages Khaling, Kangpara and other places. Pastures are allotted *thram* (registration) numbers. Two to six households share a patch of grazing land with a particular *thram* number, while each household pays an individual annual tax of Nu.100 to the Government. Only about 45 households (28.72%) own grazing lands individually in Merak. The holding size ranges from 10 acres to about 810 acres for the summer pastures. It ranges from a minimum of 8 acres to a maximum of 180 acres in autumn pastures and between 13 to 400 acres in the winter pastures. It is noteworthy that 63.96% of the total pasturelands in Merak is found to be located in the summer high altitude areas. Out of total of 157 households in Merak, about 39.4%, 19.10% and 15.92% households have no pasturelands in autumn, winter and summer pastures respectively. Most of the households owing autumn and winter pastures have pasture holding less than 100 acres. Only 13.36% and 5.08% households own more than 100 acres in the above two categories of pastures.

3.4.5.1 Pasture Management

The pasture development practices and sowing of improved fodder seeds were started from 1987 (RNR Newsletter, 1999, p. 8). Most of the farmers have sown the improved fodder seeds in their *tsamdrog* and few have even sown them in their kitchen gardens. The grasses in the improved pasturelands are white clover, cook's foot, Italian ryegrass, Maculates (legume) and Talfescue. The area under hybrid grasses and herders adopting this practice in Merak is not very significant. However, the herders establish these new grasses by broadcasting the fodder seeds in the open ground close to their camps and in areas where bamboos, shrubs and bushes have been cut and burned in the previous winter. A maximum of 41 herders from Merak cultivated hybrid grasses in 41.41 acres of pastureland in 1994–1995. This practice declined as only 25 households sowed hybrid grasses in 5 acres of land in 1998–1999. One of the main reasons for improved pastures not being taken seriously may be the absence of the practice of keeping calves in the sheds overnight like other parts of Bhutan. Secondly, when the herds migrate in autumn to these pastures, the thickly grown improved grasses are devoured within a day or two due to uncontrolled grazing. For the rest of the winter, the improved pasture is nothing more than a badly treated grassy area and, in the summer, when the herds are moved to a higher altitude, the pastures are grazed either by stray horses, cattle or wild animals. The adoption of improved pasture may be high but management is poor and not taken seriously (RNR Newsletter, 1999, p. 9). A number of trials and experiments have been done to overcome the problem of overgrazing in the area. Brokpas have lately shown more interest on the improved pasture management than the fodder trees and crop residues.

3.4.5.2 Transhumance and Drukor

In the truest sense of the word, the Brokpas cannot be identified as nomads. It would be more correct to call them transhumant as they have a well-defined seasonal cycle of movements. The migratory pattern of their life is an economic necessity. The extreme climate of the place makes it hard to live there throughout the year. As the cold weather sets in, entire families have to move their herds across several passes, dropping lower to the warmer regions for frost-free pastures. They move slowly over a period of 4–6 weeks along with their cattle down up to 2500 m in different directions. The Brokpas of Merak follow four main routes. The first group marches towards Khilphu and the second towards Shingkhar along the river Jomori in the south. The third group follows Nyeramachhu valley and then Brongzongla ridge up to Danglingtsho diverting into two main branches: one to Kangpara and the other towards Khaling and Wamrong. The fourth group roams around Mindula pass between Galing and Phongme villages.

The winter pasture movement of the people of Sakteng breaks up into two directions. Some of them go up to Shingkhar and Khilphu crossing the Nyakchungla pass. The majority of them move down along Gamrichhu on both sides. The summer movement of the Sakteng people is confined to Zangphu Sum range bordering Arunachal Pradhesh of India, while the people of Merak go up to Sherphu, Jomokunkhar, Dangzong and the region around the Nyakchungla pass.

While this migration is a climatic compulsion and an example of the ecological adaptation of the Brokpas it also provides them the opportunities to go for *drukor* during winters. *'Drukor' or 'brukor'* roughly means moving around for grains to exchange with their dairy products. *Drukor* is thus a unique trading relationship that exists between Brokpas and their Sharchop hosts and friends. These strong social ties bind together the highlanders and the valley dwellers underlying their economic necessities. While some men and children tend the yaks in nearby pastures, other members of the family come down to the villages and barter their products.

Almost all the Brokpa families have *`nepos'* i.e. village hosts. In summer months the Brokpas bring down butter and cheese and their *nepos* sell it for them in the village for them in exchange for rice, maize and chillies. The *drukor* practice starts by late October and continues for 2 or 3 months. They move from one village to another collecting grains that have been bartered for butter and cheese by their *nepos.* The village hosts also help in the Brokpas in gathering their grains from various families. It is imperative for a Brokpa to eat and lodge at his *nepo's* house during his stay in the village. The *nepo* in turn receives cheese and butter as gift. A single Brokpa family will have one or more *nepo* families. This tradition has been sustained through several generations.

The Brokpas from Sakteng usually trade with Tawang and Bomdilla regions of Arunachal Pradhesh in India. Brokpas still trade with their Monpa *nepos* across the border and visit as far as Udalguri in Assam to buy raw silk, cooking pots, salt, gumboots etc. The *Drukor* tradition and practice is still very much an essential economic ingredient for the Brokpa society. According to an estimate, one Brokpa family traded close to 300 kg of quintals of butter and 60 kg of cheese and *yitpcz.* In exchange, the family received 35 quintals of un-husked rice and 30 quintals of maize grains in the winter of 1999. Another family had bartered two hand woven blankets for 280 *changdeys* (one *changdey is* about 2 kg) of maize and dried chilly. About 14 *changdeys* of maize was bartered for 12 kg of butter. In cash, one blanket is sold for Nu.5000, butter for Nu.120 per kg and *yitpa* for Nu.90 per kg. Some Brokpas buy rice from government subsidized fare priced shops and prefer to sell butter and cheese in cash. The *Gup* (Headman)of Merak was, however, of the opinion that the un-husked rice, which they receive from the villagers is less profitable and demands more labour but as they need maize to feed the herds so they need to do the *drukor.* Increased access to market centres facilitates monetary transaction practices. One dissatisfying factor to this effect has been that urban population does not seem to like the butter and cheese prepared by the Brokpas. *Drukor* too has lost its original significance and value that was once attached to it. Nevertheless, the *drukor* tradition is still essential to the Brokpas, at least as long as they continue with yak herding and occupy their original homes in the upper reaches of the mountains.

References

Ali, M.Y., Sarower, M.G., Banu, G.R., & Rouf, M.A. (2019). Identifying peak breeding season of mud crab: A timely Intiative. *A Scientific Brochure Published by CARE-Bangladesh for Strengthkening the Capacity of Crab Hatchery and Local Service Providers under Shomoshti.* Project 2019, pp. 1–6.

Dey, D. (2002). Bhutan's current strategy on environmental management. *International Review for Environmental Strategies IGES, 3*(2), 20–30.

Dey, D (2010). *Conserving wetlands through microfinance programs, India.* TEEB case study; http://www.teebweb.org/wp-content/uploads/2013/01/Conserving-wetlands-through-microfinance-programs-India.pdf

Dey & Pankaj (2003). *Proceedings of international UNFPA symposium on population and environment, Thimphu, Bhutan 2003* (pp. 44–51).

Dey, D., & Sadruddin, (2000). Biodiversity in Bhutan Himalayas: status and conservation of forests in 2000. *Proceedings of the International Congress on Forests, BIOSFERA* (pp. 156–167). Brazil.

Dorji, T. (1995). *Assessment of non-timber forest resources utilization in eastern Bhutan.* Field Doc. No.8, FSD, MoA, Govt. of Bhutan.

Fukuyama, F. (1998). *Trust: The social virtues and creation of prosperity.* The Free Press.

Gilmour, D. A., & Fisher, R. J. (1996). *Villagers, forests and foresters of Himalaya.* Sahayogi Press.

Govt. of Bhutan. (1997). 8th F.Y. Plan Document, Mo Planning.

Grierson, A. J. C., & Long, D. G. (1983). Flora of Bhutan. *Edinburgh: Royal Botanical Garde, 1*(1), 143–155.

Gyamtsho, P. (1998). *Status of biodiversity conservation in Bhutan.* In Report on the International Meeting on Himalaya Eco-Regional Cooperation. (11–52). UNDP,WWF& ICIMOD.

IUCN. (1997). *A global overview of wetland and marine protected areas on the world heritage list: A contribution to the global theme study of world heritage natural sites, prepared by natural heritage Programme.* IUCN, Gland, Switzerland in collaboration with the World Conservation Monitoring Centre, Sept. 1997.

Kirby, J (2010). *Review of Current knowledge of bird flyways, principal knowledge gaps and conservation priorities.* UNEP/CMS/ScC16/Doc.10 Annex 2b; https://www.cms.int/sites/default/files/document/ScC16_Doc_10_Annex_2b_Flyway_WG_Review2_Report_Eonly_0.pdf

Lobzang. (1998). Biodiversity in Bhutan: a conservationist's view. *Academic Journal of Sherubtse College, 4*(1,2), 112–116.

Messerschimdt, D. Chhetri, B., & Pradhan, M. (1996). *Upper slope forest management in Kabhre Palanchok, Nepal.* Consultancy report on the Nepal Australia community forestry development project. ICIMOD Kathmandu.

Messerschimdt, D. Dorji, D. & Thinley, K. (2000). *Social forestry in eastern Bhutan. 3rd forestry development project. RNR-RC, Khangma.* Bhutan.

Namgyel, P. (1996). *Beyond timber: What value of forests*. Country Paper. Bhutan, Yasupang: Forest Research Institute, MoA, Govt. of Bhutan.

Norbu, L. (1994). *Forest tree improvement in Bhutan: A base line study*. FAO, working paper No. 5.

Ramsar Secretariat. (2010). http://www.ramsar.org/about/about_biodiversity.htm

SAEFL. (2004). *Living waters, conserving the source of life: The economic values of the World's wetlands*. Prepared with support from the Swiss Agency for the Environment, Forests and Landscape (SAEFL), gland/Amsterdam, January 2004.

Vyas, P. (2005, October 25–October 28). *Human-Tiger conflict resolution in Sunderban Tiger Reserve (India)- current management practices and strategy for long term conservation*. Proceedings of Fifth International Workshop on 'Ecotourism and Biodiversity: Shrinking Wetlands' organized by the Indo-American Environmental Leadership Programme of United States Educational Foundation in India in association with Forest Department, Govt of West Bengal held at Sunderban, West Bengal.

Wasi, P. (1997). Community forestry, the great integrative force. In M. C. Victor et al. (Eds.), *Community forestry at a crossroads* (pp. 3–8).

WWF. (1999). *A manual for wildlife management in Bhutan* (World Wildlife Fund, Bhutan Chapter, MoAg). Thimphu.

Chapter 4
Biorights as a Financial Mechanism

R. Gopichandran and Dipayan Dey

4.1 Facets of Biorights as a Financial Mechanism

The spread and depth of deliberations on the framework of ecosystem services especially at the interface of biorights and sustaining transitions has grown over the past decade and a half. This is especially so, with the call for integrated mitigation and adaptation outcomes, based on growing evidence of artificially induced variations in climate and challenges posed by such perturbations seamlessly across geographies. The centrality of ecosystems services and imperatives to sustain their ameliorative functions cannot be overemphasized. Links between environmental and economic perspectives of climate resilient transitions could modulate patterns of consumption and use of goods and services derived from ecosystems and income generating activities for livelihood of beneficiaries. Markets tend to distribute resources as a function of demand and supply of a material or service. The market values of such ecosystem goods and services help understand qualitative and quantitative dynamics of such an engagement. Their precise forms and functions of these correlates however continue to remain elusive.

The exchange values of services present the economic perspective, and ecologists focus on expenses incurred in maintaining the health and resilience of the ecosystem concomitant with uninterrupted delivery of such services (Bingham et al., 1995). Sociological values are also attributed to its moral assessments thereof (Barry & Oelschlaeger, 1996; Cornes & Sandler, 1996). MEA (2003) defined the

R. Gopichandran (✉)
NTPC School of Business, Noida, India

NCR, Delhi, India
e-mail: gopichandran@nsb.ac.in

D. Dey
South Asian Forum for Environment, Kolkata, West Bengal, India

D. Dey, J. Roy (eds.), *Biorights*, Studies in Ecological Economics 7,
https://doi.org/10.1007/978-3-030-91503-2_4

value as the "contribution of an action or object to user specified goals, objectives or conditions" Nature based solutions have been at the centre of GEF's initiatives to deliver mitigation and adaptation outcomes reinforced by Leaders Pledge for Nature. The Scientific, Technical and Advisory Panel called for a system – approach to expedite green and blue transitions (UNFCCC, 2021). The Global Commission on Adaptation GCA & WRI (2021) emphasized on the need for windows to harvest and embed private sector investment. Businesses that promote use of alternative energy sources could enrich the portfolio of projects.

Guerry et al. (2015) invited the attention of development agencies to deliberate on logical frameworks to incorporate dynamics of natural capital and ecosystem services in decision-making. This followed the impetus provided by the Millennium Ecosystem Assessment and continuing disparities due to inadequate evidence-based planning, and the incongruence between stakeholders related often short-term outlooks compared to the much-needed societal long-term goals (Guerry et al., 2015). Mengist et al. (2020) deliberated on regulatory ecosystem services as an overarching framework. The chemical ecological basis of such transitions is equally important in this context (Addison et al., 2018).

Donatti et al. (2019) present Indicators to measure the climate change adaptation outcomes of ecosystem-based adaptation and indicted that consensus about ecosystem – based adaptation successes are yet emerging for want of a convergent template of systems – specific leads. They present a matrix of several parameters with a special call for long term planning perspectives. The World Economic Forum recently presented the Circular Economy Handbook (Jensen, 2022) to help optimize through" the born-circular's" including improved access to products and services.

While ecosystem goods, e.g., crops, timber etc. have definite market values, there are typical challenges of internalizing ecosystem services into market systems. This is largely because direct and non-direct values of these services are not adequately aligned with market economy. Appropriately designed market instruments are needed to foster conservation action and outcomes at lesser costs and simultaneously create positive incentives to intensify ecosystem services. Innovations and system augments can complement such initiatives (Stavins, 2003). Such system – specific financial mechanisms may induce much needed behavioural changes in communities and local institutions to use nature's goods and services in cost-effective ways. The Convention on Biological diversity, especially after the 2010 Nagoya meet, Ramsar Convention on Wetlands, Convention on Desertification and SDGs and Aichi Targets provide the impetus to embed environmental goods as part of national accounts. That the contours of PES are only evolving is acknowledged by the CBD too. This involves an interplay of markets for such specific services as watersheds, climate functions, and aesthetics to name a few. They impose individual and synergistic impacts on outcomes; determined also by property rights, and the timing of market instruments. The larger challenge is the persisting question about cost – optimality, aligned also with pre-existing conditions especially of differences in access of benefits between the already endowed and the real poor within communities (Mayrand & Paquin, 2004; Guerry et al., 2015).

Gaworecki (2021) argued on the basis of dozens of studies that there is significant scope to raise *payments because considering the high opportunity costs of options the land resources could have served.* Börner et al. (2017) *and also as cited by* Gaworecki (2021) *highlighted the interplay of* institutional and socio-economic conditions, environmental service values and provision costs, political enablers and levels of payments including tenures. As an optimal resource allocation system. Yet, it focuses on decision support systems for resource allocation through several institutional arrangements (Williamson, 1985; Stiglitz, 1986; North, 1990). Market mechanisms or financial arrangements are at best options for decision making in conservation and development planning, only to better engage communities in conservation. They are not overarching as a panacea for environmental problems; especially of global scale climate change. The fourth TEEB Report of UNEP highlighted the importance of biorights in incentivized community-based conservation (Dey, 2010) as an alternative to 'command and control' of community users. This harmonizes institutional arrangement with financial incentives, conservation objectives, societal values and public good. Societal values and perception of public good vary widely across space and time.

India's net – zero commitment and clarity about outcomes and approaches that target SDGs (NITI Aayog, 2021) create the setting to explore community – led augments that seamlessly secure economic and environmental gains. Her national missions address sectors of immediate production and consumption consequence, across all economic strata. Agriculture, forestry and fisheries are important priority targets from the biorights perspective involving land and water resources. The periodicity and intensity of impacts may influence the choice of mutually reinforcing conservation and remedial measures. Financial support and other institutional mechanisms should sensitize stakeholders who benefit from environmental services to pay for such services and the ones who help generate such resources be compensated adequately. The challenge is to however influence decision making processes and priorities of those who generate the externalities and are not directly affected by them. An excellent recent analysis by Jha et al. (2022) established differences across altitudes in the scale of reduction in the flow of ecosystem services.

The present framework of analysis confines natural resources to the context of biorights. The premise is the consumption of a resource by one or a group of individuals or communities; not coveted, does not affect others. The, market of such a good(s) is not demand driven and the motive to produce it in excess gradually decreases. Similarly, if any good or services are non-excludable, everyone consumes or uses it. Demand for such goods or services increases and supply is constrained by limits of capacities of ecosystems they are derived from. Such goods are not considered as traded between users and providers as there are no incentives for providers of these goods and service. Users under such circumstances take a free ride. Markets in both cases could fail due to their own limitation Institutional arrangements within the biorights framework attempt to create market dynamics with added values. Several principles of engagement appear to modulate dynamics of consumption and resultant management regimes; as stated. Interesting attributes of such circumstances are biorights centred market linked paradigm are presented.

The short-term contract and one-time grant model are about a conservation centred community-based conservation activities, promoted by a pilot in an area with acute conservation need. Ironically communities inhabiting the area do not prioritize conservation need. This is because the market value of conservation is ambiguous and the product is not competitive in nature. Some project support that enhances infrastructure or capacities is extended through a one-time grant for the beneficiaries. This inadvertently creates a local market for the product. Support is designed to co-terminate with the emergence of the market. The contract could however be extended to scale up interventions or replicate on mutual consent. These realities are reflected in the case of highlanders of terrain in eastern Himalayas.

The highlanders under reference collect wild mushrooms and medicinal plants from far removed remote valleys. These materials fetch a good market price in nearby villages or hamlets. They however are not able to sustain such price rewards because they cannot supply consistently. This is because they tend to overexploit and often fall victims to man-animal conflict and such calamities as landslides and, floods due to glacial lake outbursts. Their returns are therefore not guaranteed. A small one-time grant, to train such a community on mushroom farming and manage a low-cost greenhouse ensured supply of quality mushrooms. This also helped structure market linkages and prevented free – ride exploitation.

The long-term contract and micro-financing model deals with support for communities, including direct beneficiaries far beyond the tenure of the project. The hypothesis is that, the biorights intervention will create the economic opportunity of an adaptive alternative to compensate opportunities lost in conservation. Additionally, the beneficiaries will be supported with a micro-financing module to create a resilient market that sustains benefits. Environmental sustainability that yields non – excludable public goods is also ensured through participatory conservation. The contract with the community is indeterminate as it gets linked with the rolling of the micro-financing and intensification of market economy.

Another case example explains this arrangement further. It is about intensive fish culture in Siska Beel, a peri-urban wetland in Guwahati, Assam. All endemic fish species were lost due to habitat destruction. The production of major carps too was falling sharply. Much needed habitat restoration however imposed a huge opportunity cost for the fishers. Micro-financing of the local fisher's groups helped diversify their trade with value chains. Diversification spread fiscal risk. and regulated marketing of local fish as a low-volume high-value product expanded market opportunity. Self – recruiting fish species enabled habitat restoration in systems that were only optimally exploited.

The third variant is about a non-renewable long-term contract and revolving fund model. It is in the setting of the flood plains of river Kosi, in the State of Bihar, along India-Nepal borders. Flood water stays for 6–7 months. Marginal farmers and fishers suffer heavily during this flooded non-productive period. They are forced to temporarily migrate as casual labourers to earn livelihood from informal sectors in urban areas. Their travails were addressed through a systemic approach. They were trained in hydroponic float-farming and integrated aquaculture as well to intensify primary productivity in these floods inundated areas. After completion of their training a one-time bulk fund was determined in the bank as collateral and soft loans were disbursed against the fund with a 2-crop cycle moratorium period. The condition was that the returns as received against loan would further be invested by the bank to newer groups to replicate the model. This created a revolving fund for easy credit linkages; ensured production from the floating farms or economies of scale. The fixed amount evolved into a corpus to cover risks due to natural disasters. This could also qualify for carbon financing due to integrated adaption outcomes.

The Auto-renewable long-term contract and PES model of biorights emerges when financial compensation is leveraged from Payments of Environmental Services (PES) after the sustainable intensification of ecosystem services by its users. The contract is between the natural resource users and providers. It is auto-renewed when the conservation objectives are fulfilled. The biorights intervention as a socio-economic catalyst expedites the contract. Technology cooperation and internalizing PES in local economy are at the core of this engagement. These elements were reflected in the following case example.

Stabilization of earthen embankments in the highly vulnerable coastal areas of deltaic Sundarbans in Bangladesh and eastern India poses a big challenge for inhabitants and other stakeholders. Embankments breach during high tide and cyclones leads. Farmlands are inundated only to accentuate consequent extreme poverty and migration. These mud embankments can't be replaced with concrete or polymer-based materials. They could disrupt system – specific bio-geochemical cycles and bio – resources ecology. On the other hand, extreme climate vulnerability and poverty compels the inhabitants to extract mud crabs and fishes, as well fuel woods from mangroves. This erodes and further weakening the embankments.

Biorights can play a substantial role in freeing the communities from referred nexus. Ecosystem-based adaptation through energy plantations could reinforce embankments and resultant fodder and fuel help communities. Additional crab-fattening, seaweed culture and extraction of non-timber forest products like honey, natural fibres and dyes etc. compensates opportunity costs of the communities. Interestingly a local niche market economy also develops; reinforced with ecotourism. Therefore, the contract between the mangrove ecosystem and the service user community is auto renewed with ecosystem-based collective community actions. Services and goods generated on account of the restored embankments are neither competitive nor non-excluded in the local niche market. This enhances community disaster preparedness and climate resilience as well.

4.2 Some Additional Models in Vogue Aligned with Biorights

Financial arrangements in biorights market model are unique; yet create impacts comparable with other arrangements. The menu of prerequisites, approaches and outcomes of the different financial conservation models lends itself to interesting comparisons.

ICDP: Integrated Conservation and Development Projects Interventions are designed to twin socio-economic development and consecration through focused participation of ultra – poor communities in particular that depend on eco-systems for their livelihoods (Franks & Blomley, 2004). ICDP premised those investments to develop commercial green – enterprises also protect ecosystem goods and services and enable sustainability. This is to create market linkages in the local area; that can be manoeuvred through mutually reinforcing infrastructure in the vicinity. This concept failed reportedly because funds and durations of contact were not adequate to create or sustain changes in habits and planning outcomes. Trade-offs between biodiversity conservation, eco-system level changes and livelihoods were not addressed comprehensively.

PES: Payment of Environmental Services refers to voluntary transactions. This involves a well-defined environmental service, or a land use likely to secure the service. This in turn is 'bought' by a minimum of one service buyer from a minimum of one service provider, only if the service provider secures service provision conditionality!!! conditionally (Wunder, 2005). These include examples of (Eijk & Kumar, 2009) of water shed management in Bolivia that reward farmers through public funds (Asquith et al., 2008) and extraction of mineral water in France by a business enterprise (Perrot- Maître, 2006), involving a spectrum of stakeholders. These relate also with contexts where users of ecosystem services themselves fund for the services. They derive advantages entirely by which the PES financing mechanism becomes a closed domain unto them.

The Centre for Science and Environment recently drew attention to a State – driven dynamic; and those deeper insights are needed to monetize nature's services. It cited ten Himalayan states asking for incentives to keep ecosystems robust, in addition to the support they received for action on forest cover. There are some estimates of the quantum of services such ecosystems provide (Down To Earth, 2019) Importantly PES may not be the most appropriate approach universally. This is because convergence is not always assured between conservation targets and market tools. Threats that deplete resources and variable opportunity costs may compound the contrast further. ESCAP highlights such useful opportunities as improved land use planning, pricing mechanisms, rights and tenures, investments in infrastructure for better delivery of remedial and preventive strategies (UNESCAP, 2009).

Public financed PES typically involves voluntarily funded interventions to sustain one or a group of ecosystem services for collective use. However, unregulated usages and irresponsible consumption of natural resources tend to derail such a coming together. The failure of market mechanism based on payments of environmental services has been critically reviewed for its defects. Often payments were tied to a proxy rather than quantifiable measured ecosystem function or process. For example, the services of a reclaimed wetland that supports ecotourism earns revenues from the proxy markets of tourism. These revenues are not equated with the service value since the pricing of the services are not system – specific. Their determinants are not deciphered adequately and therefore It also is known to generate perverse incentives in some contexts (Tattenbach et al., 2006) or poverty alleviation was not adequately addressed (Pagiola & Platais, 2006).

PES interventions in many cases had small gains over much higher opportunity cost that was incurred by the marginal communities. Such low incentives failed to motivate the community for partnering in conservation activities in many geographies The PES approach addresses the classic problem of environmental externalities (Wunder, 2005; Richards & Jenkins, 2007; UNESCAP, 2009). It is however essential to sustain the service through appropriate conservation and remedial measures.

4.3 REDD, REDD+ and UN-REDD: Reduction of Emission from Deforestation and Degradation of Forest

It is well known that emissions on account of forest degradation and deforestation are next only to that of the power sector. The REDD aspired to create a monetary value for the stored carbon in forests. This was to incentivize reduction of emissions from forested lands and the same could be invested in low-carbon or carbon-neutral pathways. REDD+ addressed conservation of forests through community-based management to enhance forest carbon stock. This was to also make indiscriminate logging less profitable. A trust fund could also emerge for the benefit of forests and serve as a cheaper option than carbon capture or use of renewable energy systems. Some challenges however were about preventing leakages, precise quantification of emission reduction and rightful resource allocation for beneficiaries, equity and reciprocity. External market pressures may further weaken their land and resource right and food security., Some sporadic successes of REDD are reported from different parts of the world. The Juma Sustainable Development Reserve Project in Amazonas, Brazil could be successfully and independently validated. It benefitted hundreds of families in threatened conservation areas, through bank linkages; covering thousands of hectares of amazon forest.

4.4 Blue Carbon

That ocean and such associated costal ecosystems as tidal wetlands, mangroves, estuaries, seaweed fields, salt marshes etc. modulate carbon dynamics is well known (Keppler et al., 2020; Hauck et al., 2020; Bourgeois et al., 2022). They reduce vulnerability to disasters alongside; provide an array of ecosystem services; important for sustainable livelihood of marginal coastal communities.

Rewards from carbon trade similar to coastal wetlands contexts can incentivize conservation by poor coastal communities. Australia's Emission reduction Fund addresses opportunities in related niches in addition to the scope defined by the

IUCN and the WEF recently (Australian Government, Department of Industry, Science, Energy and Resources, 2022; Favasuli, 2021; IUCN, 2021; Johnston, 2021).

Comparing the attributes of the financing mechanisms as discussed above, we derive a 'Heat Map' of multi-criteria decision support tools. These attributes in each of the financing mechanisms are weighed in a percentile scale and scores are placed in respective cells of the heat map. The weightages are arbitrarily attributed and the and colour index in the cells of the heat map grades from 'no-fill' to 'deep-fill' with increasing values.

Program Name	ICDP	PES	Micro-Fin	REDD	Blue Carb	Biorights
Program Area	Any	ESs	Any	Forest	Coastal	Any
Dev Financing	25	15	35	15	5	5
Conv. Financing	15	20	0	20	20	25
Stakeholders Part.	10	15	5	20	15	35
Poverty Alleviation	5	15	25	20	10	25
Capacity Building	10	10	5	25	15	35
Mon & Evaluation	5	10	15	20	15	35
Impact Assessment	5	15	0	20	25	35
TOTAL SCORES	**75**	**100**	**85**	**140**	**105**	**195**

The Critical Ecosystem Partnership Fund (CEPF, 2022) provides an interesting landscape of pillars and cross cutting elements as substantiated also by the FAO (FAO, 2022). Olander et al. (2018) referred to the growing inclusion of ecosystem on the basis that inaction could generate other consequences. The Intergovernmental Science Policy Platform on Biodiversity and Ecosystem Services (IPBES, 2020) and the Natural Capital Declaration's (Capitals Coalition, 2021a) natural capital protocol (Capitals Coalition, 2022) and the social and human capital protocol enrich our understanding of these dynamics (Capitals Coalition, 2021b). along with benefit relevant indicators and "final or intermediate ecosystem goods and services" (FEGs) as they prescribe. They emphasize the need for standards for ecosystem service measures.

4.5 Other Emerging Contours

Eijk & Kumar (2009) presented important elements of the biorights framework with case examples from many geographies. They superimpose on the management of transaction costs and leakages in particular. In case of PES or ICDP, this pertains to opportunity costs incurred by communities for benefits forgone, implementation costs, capacity building, monitoring and evaluation. In case of a micro-finance, time costs and levels of financial literacy also matter. In case of biorights, costs could be higher because of assessments of poverty alleviation and balancing equity. Insights about the latter two determine contours of community engagement. Costs could reduce when communities are trained to and co – monitor and evaluate outcomes and materials are refurbished and reused in amelioration techniques (Eijk & Kumar, 2009).

Sub-optimal interventions that do not address the root cause of challenges could create and enhance leakages through avoidable dislocation, damage and decay. This could be true in the management of pasturelands and water bodies in particular. Stakeholder partnership is at the centre of the biorights paradigm. Multiple overlaps of legal and administrative jurisdiction of holders of property and governance authorities could affect consensus, participation, responsibility and equity outcomes. Wunder et al. (2008) highlight eligibility, desire, ability and competitiveness of the stakeholders; essential to even map opportunities scale and tenure of interventions with respect to defined baselines. Perverse incentives should be avoided to prevent camouflaged interests. This came to light when farmers deliberately reported less fish yields in order to engage in eco-tourism. The latter was actually an alternative to only tide over challenging times; while sustainable fishing itself could be restored with eco-friendly practices.

Self-regulatory feedback mechanism and circularity in biorights design makes it more appropriate in that way. One of the most recent examples of community led mitigation and adaptation in peat and mangrove systems highlights these dynamics (IKI, 2022).

4.6 Building Capacities for Sustainable Ecosystem Services

Changes in the qualitative and quantitative profiles of resources could be due to natural or artificially induced perturbations and cascading impacts. Climate change related impacts further compound outcomes of variations. Chemical ecological investigations on age-correlated changes in the biochemical profiles of agriculture and forestry vegetation reveal these correlates. They influence the occurrence and distribution of such related bio resources as natural enemies of pests and microbes that determine nutrient cycling in soils. This could be true even of the ecology of fishes and other aquatic bio resources. Recent emphasis on heat shock proteins (Cho et al., 2021) and marine heat waves (Virginia Institute of Marine Science, 2022; Tanaka & van Houtan, 2022) & call for holistic management strategies.

Communities are well known for their abilities to innovate measures to conserve natural resources aligned with the system – specific physical, chemical and biological characteristics; with implications for sustaining ecosystem services (World Bank Group, 2020; Corbera et al., 2009; Wendland et al., 2010; Llambi & Morra, 2010; Pirard et al., 2010; Tacconi et al., 2011; WRI, 2009; World Development Report of World Bank, 2010; SGP India, 2022; SRISTI, 2021). Institutional mechanisms that enable conservation and prevent communities from such over exploitation are also seen in many cases (Navarro, 2010). Gaps in knowledge regarding mitigation and adaptation dynamics may retard the development and implementation of preventive management practices. It is important to therefore ask if adequate attention is being paid to the ecological aspects of ecosystem services and draw lessons from environmental management protocols that have successfully addressed issues relating the use of alternatives.

4.7 Towards Sustaining Ecosystem Services

The GEF in a recent civil society consultation through its GEFCSO network called to redouble conservation and economic gains. This took note of the outcomes of scores of community-based climate leadership programmes including PES (GEF, 2021a, b) The Special Climate Change Fund invited attention to paucity of abilities to foresee and manage climate risks, and locally adapted technologies and hence the scope for business enterprises and markets to assist transitions (GEF, 2021a; Wagle, 2017). This is aligned with the Business Call to Action of the UNDP that attracted the attention of hundreds of stakeholders' world over (UNDP, 2018). UNDP's Small Grants Programme through its vulnerability reduction assessment and community-based adaptation approaches reiterated these insights (UNDP, 2021a, b). The Climate Resilience and Adaptation Finance and Technology Transfer Facility (CRAFT) that created the opportunity for companies to assist resilience interventions (GEF, 2017). This complements the groundswell created by the GEF Challenge Programme for Adaptation Innovation (GEF, 2019) with opportunities to learn from IIED's Global Initiative on Community – Based Adaptation (IIED, 2018). One of the latest leads in assessing adaptation opportunities comes from Tschanz et al. (2022) with spatially and temporally significant socio-ecological indicators.

India's National Adaptation Fund for Climate Change (NAFCC) established in 2015 such locally indispensable adaptation interventions as groundwater recharge with positive implications for sustainable, livestock, agriculture and coastal habitats management. The multilateral Himalayan Climate Change Adaptation Programme HICAP over 2011–2017; developed small-scale climate and hydrology scenarios, to influence community-based decisions about ecosystem services. This relates to disaster risk reduction strategies for humans and related livestock and agriculture resources that are not amenable to physical transfers as part of disaster mitigation (MOEF & CC, 2014; TTP, 2021). Karki et al. (2021) reviewed scores of adaptation projects in the just – passed decade pertaining to Ecosystem-based Adaptations and Ecosystems Protecting Infrastructure and Communities (EPIC) in Nepal. They presented a matrix of criteria and indicators that related institutional and behavioral responses with use of technologies, plans and practices.

Scarano (2017) demonstrated that ecosystems - based adaptation triggers adaptive transitions when socio – ecological systems are suitably embedded in decision making with implications for scalability of interventions and that 17 major perspectives could determine outlooks to sustain impacts of conservation. A more recent climate – smart value approach has been analyzed by Rosenstock et al. (2020) that integrates, productivity, resilience and mitigation outcomes. These integrate concerns and capabilities of small holder farmers in particular to participate in value chains, enable and de-risk investments and assist the vulnerable in particular with appropriate products and services.

The recent EU work programme (EN, 2021) on food, bio-economy, natural resources, agriculture and environment. This integrates societal values and behaviours, production and consumption patterns, human population dynamics and trends and trade. The need for disruptive change that also ensures equity and justice through the energy environment nexus has been highlighted by the World Economic Forum (World Economic Forum, 2020). Useful lessons could also be derived from the Biodiversity Finance Initiative of the UNDP (The Biodiversity Finance Initiative, 2016) and of Nkonya (2021) focusing on lands and wilderness. The OECD recently presented a menu of actions for business engagement (OECD, 2019) through opportunities for cost – effective management of bio – resources and resultant ecosystems services.

FAO a decade ago (2010) and Pagiola (2008) reviewed features of the payments for environmental services and compensation for environmental services. It emphasized links between incentives across upstream and downstream further determined by private and basic rights; considering the Andes. Earlier in 2005 Wunder indicated that integrated conservation and development projects could simultaneously increase incomes and conserve environment duly integrating trade-offs. It is likely that ownership and quantum of resource owned determine the extent to which communities participate in payment programmes. Interestingly caps on the resource extraction are also proposed along with payments for conservation.

The Climate Witness Programme of the WWF (WWF, 2007) is an interesting bottom-up approach to compile and interpret empirical insights on locally relevant variations and guide well informed preventive and remedial action. They relate also to the insights the Eco Systems Livelihoods Adaptation Network (ELAN, 2013) the BESSA: Building Ecosystem Services Research Capacity in Semi-Arid Africa (Paulsen & Dauber, 2020) the Swedbio (SwedBio, 2022) and of course the recent thrust through NDC and the biennial reports of nations. It will be obvious from the above that a large number of initiatives are in progress and certainly involve the stakeholders in a significant manner.

The 'for people and planet' strategy relevant for the next 3 years to start with and UNDP's strategic plan for the same period (UNEP, 2021) create the necessary impetus to address the stated gaps in a concerted manner. This is signified also by the PANORAMA of the IUCN and the UNDP (IUCN, UNDP, 2014) with special reference to protected and conserved areas. Deep insights are emerging even in the specific Indian context through UNDP's community based natural resource management in India (UNDP, 2013) that builds on principles and practice of traditional knowledge systems. While these provide hope for immediate, medium- and long-term benefits for communities, the challenge of market integration has to be addressed in greater detail. A bottom – up stakeholder centred natural resources management approach is central to fulfil recently articulated net – zero commitments post – Glasgow. Citizen science practices that validate observations and insights can enhance their stake in these processes.

References

Addison, P. F. E., Bull, J. W., & Milner-Gulland, E. J. (2018). Using conservation science to advance corporate biodiversity accountability. *Conservation Biology, 33*(2), 307–318. https://doi.org/10.1111/cobi.13190

Asquith, N. M., Vargas, M. T., & Wunder, S. (2008). Selling two environmental services: In-kind payments for bird habitat and watershed protection in Los Negros, Bolivia. *Ecological Economics, 65*(4), 675–684. https://doi.org/10.1016/j.ecolecon.2007.12.014

Australian Government, Department of Industry, Science, Energy and Resources. (2022). *Tidal restoration of blue carbon ecosystems method.* Https://Www.Industry.Gov.Au/Regulations-and-Standards/Methods-for-the-Emissions-Reduction-Fund/Tidal-Restoration-of-Blue-Carbon-Ecosystems-Method. Retrieved February 13, 2022, from http://www.cleanenergyregulator.gov.au/ERF/Pages/Choosing%20a%20project%20type/Opportunities%20for%20the%20land%20sector/Vegetation%20and%20sequestration%20methods/Tidal-restoration-of-blue-carbon-ecosystems-method.aspx

Barry, D., & Oelschlaeger, M. (1996). A science for survival: Values and conservation biology. *Conservation Biology, 10*(3), 905–911. https://doi.org/10.1046/j.1523-1739.1996.10030904-2.x

Bingham, G., Bishop, R., Brody, M., Bromley, D., Clark, E. T., Cooper, W., Costanza, R., Hale, T., Hayden, G., Kellert, S., Norgaard, R., Norton, B., Payne, J., Russell, C., & Suter, G. (1995). Issues in ecosystem valuation: Improving information for decision making. *Ecological Economics, 14*(2), 73–90. https://doi.org/10.1016/0921-8009(95)00021-z

Börner, J., Baylis, K., Corbera, E., Ezzine-de-Blas, D., Honey-Rosés, J., Persson, U. M., & Wunder, S. (2017). The effectiveness of payments for environmental services. *World Development, 96*, 359–374. https://doi.org/10.1016/j.worlddev.2017.03.020

Bourgeois, T., Goris, N., Schwinger, J., & Tjiputra, J. F. (2022). Stratification constrains future heat and carbon uptake in the Southern Ocean between 30°S and 55°S. *nature. Communications, 13*(1). https://doi.org/10.1038/s41467-022-27979-5

Capitals Coalition. (2021a, August 24). *The Capitals Coalition – redefining value to transform decision making.* Capitals Coalition. Retrieved February 13, 2022, from https://capitalscoalition.org/

Capitals Coalition. (2021b, December 8). *Social & human capital protocol.* Retrieved February 13, 2022, from https://capitalscoalition.org/capitals-approach/social-human-capital-protocol/

Capitals Coalition. (2022, February 11). *Natural capital protocol.* Retrieved February 13, 2022, from https://capitalscoalition.org/capitals-approach/natural-capital-protocol/?fwp_filter_tabs=training_material

CEPF. (2022). *Monitoring and evaluation|CEPF.* CEPF Monitoring and Evaluation. Retrieved February 13, 2022, from https://www.cepf.net/impact/monitoring-and-evaluation

Cho, Y. S., Jeong, T. H., Choi, M. J., Kim, J. M., & Lim, H. K. (2021). Heat shock protein 70 gene expression and stress response of red-spotted (Epinephelus akaara) and hybrid (E. akaara female × E. lanceolatus male) groupers to heat and cold shock exposure. *Fish Physiology and Biochemistry, 47*(6), 2067–2080. https://doi.org/10.1007/s10695-021-00966-1

Corbera, E., Soberanis, C. G., & Brown, K. (2009). Institutional dimensions of payments for ecosystem services: An analysis of Mexico's carbon forestry programme. *Ecological Economics, 68*(3), 743–761. https://doi.org/10.1016/j.ecolecon.2008.06.008

Cornes, R., & Sandler, T. (1996). *The theory of externalities, public goods, and club goods* (2nd ed.). Cambridge University Press. https://doi.org/10.1017/CBO9781139174312

Dey, D. (2010, October). *Conserving Wetlands through microfinance programs India.* TEEBweb.Org. Retrieved February 12, 2022, from https://www.scribd.com/document/259263908/Conserving-Wetlands-Through-Microfinance-Programs-India

Donatti, C. I., Harvey, C. A., Hole, D., Panfil, S. N., & Schurman, H. (2019). Indicators to measure the climate change adaptation outcomes of ecosystem-based adaptation. *Climatic Change, 158*(3–4), 413–433. https://doi.org/10.1007/s10584-019-02565-9

Down To Earth. (2019, July 27). *Court digest: Major environment hearings of the week (July 22–26)*. DownToEarth. Retrieved February 12, 2022, from https://www.downtoearth.org.in/news/environment/court-digest-major-environment-hearings-of-the-week-july-22-26%2D%2D65873

Eijk, P. V., & R. Kumar, (2009). Bio-rights in theory and practice. *A financing mechanism for linking poverty alleviation and environmental conservation*. Wetlands International, Wageningen, The Netherlands. Published by Wetlands International.

ELAN. (2013, November). *Integrating community and ecosystem-based approaches in climate change adaptation responses*. https://d2ouvy59p0dg6k.cloudfront.net/downloads/integrating_community_and_ecosystem_based_approaches_in_climate_change_adaptation_res.pdf. Accessed 12 February 2022.

EN. (2021, December). *Horizon Europe – Work programme 2021–2022 food, bioeconomy, natural resources, agriculture and environment*. European Commission. https://ec.europa.eu/info/funding-tenders/opportunities/docs/2021-2027/horizon/wp-call/2021-2022/wp-9-food-bioeconomy-natural-resources-agriculture-and-environment_horizon-2021-2022_en.pdf. Accessed 12 February 2022.

FAO. (2022). *Ecosystem services & biodiversity (ESB)*. Food and Agriculture Organization of the United Nations. Retrieved February 13, 2022, from https://www.fao.org/ecosystem-services-biodiversity/en/

Favasuli, S. (2021, October 25). *"Blue carbon" gains interest in effort against greenhouse gases, but challenges remain*. S & P Global Platts. Retrieved February 13, 2022, from https://www.spglobal.com/platts/en/market-insights/latest-news/energy-transition/102521-blue-carbon-gains-interest-in-effort-against-greenhouse-gases-but-challenges-remain

Franks, P., & Blomley, T. (2004). Fitting ICD into a project framework: A care perspective. In T. O. McShane & M. P. Wells (Eds.), *Getting biodiversity projects to work* (Vol. 5, pp. 77–97). Columbia University Press. https://doi.org/10.7312/mcsh12764-007

Gaworecki, M. (2021, May 14). *Cash for conservation: Do payments for ecosystem services work?* Mongabay Environmental News. Retrieved February 12, 2022, from https://news.mongabay.com/2017/10/cash-for-conservation-do-payments-for-ecosystem-services-work/

GEF. (2017, November 13). *GEF joins forces with partners to promote new fund for resilience in the poorest countries*. Global Environment Facility. Retrieved February 13, 2022, from https://www.thegef.org/newsroom/press-releases/gef-joins-forces-partners-promote-new-fund-resilience-poorest-countries

GEF. (2019, December 10). *Winners of GEF challenge program for adaptation innovation announced*. Global Environment Facility. Retrieved February 13, 2022, from https://www.thegef.org/newsroom/press-releases/winners-gef-challenge-program-adaptation-innovation-announced#:%7E:text=The%20GEF%20Challenge%20Program%20for,changing%20weather%20and%20water%20patterns.

GEF. (2021a, June 11). *Adapting to climate change from the ground up*. Global Environment Facility. Retrieved February 13, 2022, from https://www.thegef.org/newsroom/news/adapting-climate-change-ground

GEF. (2021b). S*pecial climate change fund – SCCF*. Global Environment Facility. Retrieved February 13, 2022, from https://www.thegef.org/what-we-do/topics/special-climate-change-fund-sccf

Global Commission on Adaptation & World Resources Institute. (2021). *Climate resilience*. WRI. https://www.wri.org/initiatives/global-commission-adaptation (Accessed 12 February 2022.

Guerry, A. D., Polasky, S., Lubchenco, J., Chaplin-Kramer, R., Daily, G. C., Griffin, R., Ruckelshaus, M., Bateman, I. J., Duraiappah, A., Elmqvist, T., Feldman, M. W., Folke, C., Hoekstra, J., Kareiva, P. M., Keeler, B. L., Li, S., McKenzie, E., Ouyang, Z., Reyers, B., … Vira, B. (2015). Natural capital and ecosystem services informing decisions: From promise to practice. *Proceedings of the National Academy of Sciences, 112*(24), 7348–7355. https://doi.org/10.1073/pnas.1503751112

Hauck, J., Zeising, M., le Quéré, C., Gruber, N., Bakker, D. C. E., Bopp, L., Chau, T. T. T., Gürses, Z., Ilyina, T., Landschützer, P., Lenton, A., Resplandy, L., Rödenbeck, C., Schwinger, J., & Séférian, R. (2020). Consistency and challenges in the ocean carbon sink estimate for the global carbon budget. *Frontiers in Marine Science, 7*. https://doi.org/10.3389/fmars.2020.571720

IIED. (2018, July 3). *Global initiative on community-based adaptation (GICBA)*. International Institute for Environment and Development. Retrieved February 13, 2022, from https://www.iied.org/global-initiative-community-based-adaptation-gicba

IKI. (2022, February). *Mitigation and adaptation through conservation and sustainable livelihoods in Indonesia's peat and mangrove ecosystems – Internationale Climate Initiative (IKI)*. International Climate Initiative. Retrieved February 13, 2022, from https://www.international-climate-initiative.com/en/details/project/mitigation-and-adaptation-through-conservation-and-sustainable-livelihoods-in-indonesias-peat-and-mangrove-ecosystems-19_III_105-3046

IPBES. (2020, May 7). *IPBES Home page*. IPBES Secretariat. Retrieved February 13, 2022, from https://www.ipbes.net/

IUCN. (2021, May 11). *First guidelines to design and deploy Blue Carbon projects in Europe*. Retrieved February 13, 2022, from https://www.iucn.org/news/mediterranean/202105/first-guidelines-design-and-deploy-blue-carbon-projects-europe-and-mediterranean

IUCN, UNDP. (2014, November). *Protected and conserved areas solutions|PANORAMA*. Panorama Solutions. Retrieved February 13, 2022, from https://panorama.solutions/en/portal/protected-areas/grid/bb?page=44

Jensen, H. H. (2022, January). *5 circular economy business models that offer a competitive advantage*. World Economic Forum. https://www.weforum.org/agenda/2022/01/5-circular-economy-business-models-competitive-advantage/. Accessed 12 February 2022.

Jha, S. K., Negi, A., & Alatalo, J. M. (2022). Quantification of ecosystem services providing socio-economic benefits to customary owners of natural resources in Pauri, western Himalaya. *Current Research in Environmental Sustainability, 4*, 100121. https://doi.org/10.1016/j.crsust.2021.100121

Johnston, W. (2021, December 15). *What are blue carbon credits and how to maximise their impact*. World Economic Forum. Retrieved February 13, 2022, from https://www.weforum.org/agenda/2021/09/how-to-maximise-blue-carbon-credits/

Karki, G., Bhatta, B., Devkota, N. R., Acharya, R. P., & Kunwar, R. M. (2021). Climate change adaptation (CCA) interventions and indicators in Nepal: Implications for sustainable adaptation. *Sustainability, 13*(23), 13195. https://doi.org/10.3390/su132313195

Keppler, L., Landschützer, P., Gruber, N., Lauvset, S. K., & Stemmler, I. (2020). Seasonal carbon dynamics in the near-global ocean. *Global Biogeochemical Cycles, 34*(12). https://doi.org/10.1029/2020gb006571

Llambi, L., & Morra, D. (2010). How can compensation for environmental services (CES) and payments for environmental services (PES) as financial mechanisms improve livelihoods of smallholder producers through NRM? (POLICY BRIEF 8). *FAO*. https://www.fao.org/climatechange/25201-051394e986b121a081e897690db41cb67.pdf (Accessed 12 February 2022.

Mayrand, K., & Paquin, M. (2004, September). *Payments for environmental services: A survey and assessment of current schemes*. Unisféra International Centre. https://www.cbd.int/financial/doc/cec-pes.pdf. Accessed 12 February 2022.

Mengist, W., Soromessa, T., & Feyisa, G. L. (2020). A global view of regulatory ecosystem services: Existed knowledge, trends, and research gaps. *Ecological Processes, 9*(1). https://doi.org/10.1186/s13717-020-00241-w

Millennium Ecosystem Assessment—MEA. (2003). *Ecosystems and human well-being: A framework for assessment*. Island Press.

MOEF & CC. (2014). *NAFCC portfolio highlights in India*. NABARD. https://www.nabard.org/auth/writereaddata/File/Overall%20NAFCC%20Project%20Flyer.pdf. Accessed 12 February 2022.

Navarro, G. (2010, October). *The evolution of costa rica's carbon rights* (REDDnetAffiliation: CATIE/ODI). REDD-net programme. https://www.researchgate.net/publication/303369912_The_Evolution_of_Costa_Ricas_Carbon_Rights. Accessed 12 February 2022.

Nkonya, E. (2021). *Protecting lands and wilderness*. https://www.un.org/development/desa/dpad/wp-content/uploads/sites/45/publication/WSR_2021_BG_Nkonya.pdf. Accessed 12 February 2022.

North, D. C. (1990). *Institutions, institutional change and economic performance (political economy of institutions and decisions)*. Cambridge University Press. https://doi.org/10.1017/CBO9780511808678

OECD. (2019, May 5). *Biodiversity: Finance and the economic and business case for action – OECD*. Retrieved February 13, 2022, from https://www.oecd.org/env/resources/biodiversity/biodiversity-finance-and-the-economic-and-business-case-for-action.htm

Olander, L. P., Johnston, R. J., Tallis, H., Kagan, J., Maguire, L. A., Polasky, S., Urban, D., Boyd, J., Wainger, L., & Palmer, M. (2018). Benefit relevant indicators: Ecosystem services measures that link ecological and social outcomes. *Ecological Indicators, 85*, 1262–1272. https://doi.org/10.1016/j.ecolind.2017.12.001

Pagiola, S. (2008). Can payments for environmental services help protect coastal and marine areas. *Environment Matters*, 14–17.

Pagiola, S., & Platais, G. (2006). *Payments for environmental services: From theory to practice* (No. 3994). Sustainable Agriculture and Natural Resource Management (SANREM) Knowledgebase. http://hdl.handle.net/10919/65833. Accessed 12 February 2022.

Paulsen, H. M., & Dauber, J. (Eds.). (2020). Agroecology: Can we change our food systems? Landbauforschung. *Journal of Sustainable and Organic Agricultural Systems, 70*(2), 1–177.

Perrot-Maître, D. (2006) *The Vittel payments for ecosystem services: a "perfect" PES case?* International Institute for Environment and Development. https://pubs.iied.org/sites/default/files/pdfs/migrate/G00388.pdf. Accessed 12 February 2022.

Pirard, R., Billé, R., & Sembrés, T. (2010). Upscaling payments for environmental services (PES): Critical issues. *Tropical Conservation Science, 3*(3), 249–261. https://doi.org/10.1177/194008291000300302

Reports on SDG|NITI Aayog. (2021). *National portal of India- Niti Aayog*. Retrieved February 12, 2022, from https://www.niti.gov.in/reports-sdg

Richards, M., & Jenkins, M. (2007). Potential and challenges of payments for ecosystem services from tropical forests. *Forest policy and environment programme (FPEP). Foresty Briefing, 16*, 1–8. https://citeseerx.ist.psu.edu/viewdoc/download?doi=10.1.1.616.7105&rep=rep1&type=pdf

Rosenstock, T. S., Lubberink, R., Gondwe, S., Manyise, T., & Dentoni, D. (2020). Inclusive and adaptive business models for climate-smart value creation. *Current Opinion in Environmental Sustainability, 42*, 76–81. https://doi.org/10.1016/j.cosust.2019.12.005

Scarano, F. R. (2017). Ecosystem-based adaptation to climate change: Concept, scalability and a role for conservation science. *Perspectives in Ecology and Conservation, 15*(2), 65–73. https://doi.org/10.1016/j.pecon.2017.05.003

SGP India. (2022). *SGP India*. Retrieved February 13, 2022, from https://sgpindia.org/

SRISTI. (2021, February 13). *Our products*. Retrieved February 13, 2022, from https://www.sristi.org/our-products/

Stavins, R. N. (2003). Experience with market-based environmental policy instruments. *Environmental Degradation and Institutional Responses*, 355–435. https://doi.org/10.1016/s1574-0099(03)01014-3

Stiglitz, J. E. (1986). The new development economics. *World Development, 14*(2), 257–265.

SwedBio. (2022). *SwedBio, Stockholm Resilience Centre*. Retrieved February 13, 2022, from https://swed.bio/

Tacconi, L., Mahanty, S., & Suich, H. (2011). *Payments for environmental services, forest conservation and climate change: Livelihoods in the REDD?* Edward Elgar Publishing Limited.

Tanaka, K. R., & van Houtan, K. S. (2022). The recent normalization of historical marine heat extremes. *PLOS Climate, 1*(2), e0000007. https://doi.org/10.1371/journal.pclm.0000007

Tattenbach, F., Obando, G., & Rodríguez, J. (2006). Mejora del excedente nacional del pago por servicios ambientales. Paper presented at the Workshop on Costa Rica's Experience with Payments for Environmental Services. San José, 25–26 September 2006. (15) (PDF) Practical Alternatives to Estimate Opportunity Costs of Forest Conservation. https://www.researchgate.net/publication/241751026_Practical_Alternatives_to_Estimate_Opportunity_Costs_of_Forest_Conservation. Accessed 12 February 2022.

The Biodiversity Finance Initiative. (2016). *BIOFIN workbook mobilizing resources for biodiversity and sustainable development*. UNDP. https://www.biofin.org/sites/default/files/content/publications/undp-biofin-web_0.pdf. Accessed 12 February 2022.

Tschanz, L., Arlot, M. P., Philippe, F., Vidaud, L., Morin, S., Maldonado, E., George, E., & Spiegelberger, T. (2022). A transdisciplinary method, knowledge model and management framework for climate change adaptation in mountain areas applied in the Vercors, France. *Regional Environmental Change, 22*(1). https://doi.org/10.1007/s10113-021-01862-3

TTP. (2021, September 7). *Climate change adaptation – what is it?* The Third Pole. Retrieved February 13, 2022, from https://www.thethirdpole.net/en/hub/climate-adaptation/

UNDP. (2013). *Biodiversity conservation through community-based natural resource management (2008–2012)*. UNDP in India. Retrieved February 13, 2022, from https://www.in.undp.org/content/india/en/home/operations/projects/closed/biodiversity_conservationthroughcommunitybasednaturalresourceman.html

UNDP. (2018). *Business call to action|United Nations Development Programme*. United Nations Development Programme. Retrieved February 13, 2022, from https://www.undp.org/partners/private-sector/BCTA

UNDP. (2021a). *Sustainable development goals|United Nations development programme*. Retrieved February 13, 2022, from https://www.undp.org/sustainable-development-goals?utm_source=EN&utm_medium=GSR&utm_content=US_UNDP_PaidSearch_Brand_English&utm_campaign=CENTRAL&c_src=CENTRAL&c_src2=GSR&gclid=Cj0KCQiA0p2QBhDvARIsAACSOON1QctoR5AGk01J4EHqWlsoulLkBC-a7ycaFi_UpjAHt9RvJZ-_DsIaAmlpEALw_wcB

UNDP. (2021b, September 2). *Sustainable development goals|United Nations development programme*. Retrieved February 13, 2022, from https://www.undp.org/sustainable-development-goals?utm_source=EN&utm_medium=GSR&utm_content=US_UNDP_PaidSearch_Brand_English&utm_campaign=CENTRAL&c_src=CENTRAL&c_src2=GSR&gclid=CjwKCAiA9aKQBhBREiwAyGP5lWZkN0keUV3uSBjtdpL2qtMcPKhHkHaibOEYk51cwFPKAFdEcARcZhoCpNcQAvD_BwE

UNEP. (2021, February). *Policy and strategy for people and planet: the UNEP strategy for 2022–2025*. https://www.unep.org/resources/policy-and-strategy/people-and-planet-unep-strategy-2022-2025. Accessed 12 February 2022.

UNESCAP. (2009). *Innovative socio-economic policy for improving environmental performance: Payments for ecosystem services* (ST/ESCAP/2560). United Nations Publication. https://www.unescap.org/sites/default/files/publications/payments%20for%20ecosystem%20services.pdf. Accessed 12 February 2022.

United Nations Framework Convention on Climate Change. (2021, September). *Report of the global environment facility to the twenty-sixth session of the conference of the parties to the United Nations framework convention on climate change* (FCCC/CP/2021/9). UNFCCC. https://unfccc.int/sites/default/files/resource/cp2021_09E%20.pdf. Accessed 12 February 2022.

Virginia Institute of Marine Science. (2022, January 18). *Increase in marine heat waves threatens coastal habitats: With continued warming, Chesapeake Bay will suffer extreme heat for half the year by 2100*. ScienceDaily. Retrieved February 13, 2022, from https://www.sciencedaily.com/releases/2022/01/220118125136.htm

Wagle, S. J. (2017, November 13). *The lightsmith group receives $1.6 million of funding from nordic development fund and global environment facility for climate resilience investment* *strategy*. The Lightsmith Group. Retrieved February 13, 2022, from https://lightsmithgp.com/news-posts/the-lightsmith-group-receives-1-6m-from-ndf-and-gef/

Wendland, K. J., Honzák, M., Portela, R., Vitale, B., Rubinoff, S., & Randrianarisoa, J. (2010). Targeting and implementing payments for ecosystem services: Opportunities for bundling biodiversity conservation with carbon and water services in Madagascar. *Ecological Economics, 69*(11), 2093–2107. https://doi.org/10.1016/j.ecolecon.2009.01.002

Williamson, Oliver E. (1985). *The economic institutions of capitalism: Firms, markets, relational contracting*. University of Illinois at Urbana-Champaign's Academy for Entrepreneurial Leadership Historical Research Reference in Entrepreneurship, Available at SSRN: https://ssrn.com/abstract=1496720. Accessed 12 February 2022.

World Bank. (2010). *World development report 2010: Development and climate change*. © World Bank. https://openknowledge.worldbank.org/handle/10986/4387 License: CC BY 3.0 IGO. Accessed 12 February 2022.

WorldBankGroup.(2020,November23).*TheAdaptationprinciples:6waystobuildresiliencetoclimate change*. World Bank. Retrieved February 13, 2022, from https://www.worldbank.org/en/news/feature/2020/11/17/the-adaptation-principles-6-ways-to-build-resilience-to-climate-change

World Economic Forum. (2020, January). *The future of the last-mile ecosystem*. World Economic Forum, Cologny/Geneva Switzerland. https://www3.weforum.org/docs/WEF_Future_of_the_last_mile_ecosystem.pdf. Accessed 12 February 2022.

WRI. (2009, November). *Stacking payments for ecosystem services* (f 202–729-7610). World Resources Institute. http://pdf.wri.org/factsheets/factsheet_stacking_payments_for_ecosystem_services.pdf. Accessed 12 Febrary 2022.

Wunder, S. (2005). *Payments for environmental services: Some nuts and bolts*. CIFOR. https://www.cifor.org/publications/pdf_files/OccPapers/OP-42.pdf. Accessed 12 February 2022.

Wunder, S., Engel, S., & Pagiola, S. (2008). Taking stock: A comparative analysis of payments for environmental services programs in developed and developing countries. *Ecological Economics, 65*(4), 834–852. https://doi.org/10.1016/j.ecolecon.2008.03.010

WWF. (2007, November). *Climate witness fact sheets*. https://wwf.panda.org/wwf_news/?117540/Climate-Witness-Fact-Sheets. Accessed 12 February 2022.

Chapter 5
Economics of Biorights

Dipayan Dey and Joyashree Roy

5.1 Introduction

The global narrative of climate change is transforming fast as signs of changes are becoming more perceptible. As various climate targeted interventions are showing opportunities global environmental discourse and response of developing countries are also emerging in new directions. However, there has been two way dynamics in this process of change as developing countries and indigenous communities have also been voicing their basic needs for well living (Hayward & Roy, 2019) in the climate constrained world. The dynamics of both are correlated. The reciprocal interaction between the two has also shaped the issues related to environmental conservation into global agenda towards sustainable development. The urgency for the adaptation to the Climate Change exposed vulnerable social groups/communities have been voiced for many decades by the developing countries and indigeneous communities who have been getting the burnt of increasing exposure to natural ecosystems and their livelihood for quite long time. As the exposure level is increasing in scale and among variety of communities new awakening is emerging. New opportunities are seen now in building the climate resilient societies, for the creation of sustainable enterprises and services that can address long standing poverty eradication and climate adaptation for the global south to build equity in human system and as well, conserve natural ecosystem. In a way this represents win-win strategy by turning the crisis into opportunity. The social capital in the form of cooperation

D. Dey
South Asian Forum for Environment, Kolkata, West Bengal, India

J. Roy (✉)
Asian Institute of Technology, Bangkok, Thailand

Jadavpur University, Kolkata, India
e-mail: joyashree@ait.asia

D. Dey, J. Roy (eds.), *Biorights*, Studies in Ecological Economics 7,
https://doi.org/10.1007/978-3-030-91503-2_5

and participation of the local community, many of whom are marginal, becomes essential to seize such opportunities. It is important to ensure that the livelihood of the local and marginal communities shouldn't be adversely impacted while seizing such opportunities. Distributive justice demands the marginalised communities, who are struggling with socioeconomic challenges in the agro-farming sector in both production and supply chains, as well in service delivery, need to be appropriately compensated through social protection mechanisms for unintended costs incurred by them in going green. On the contrary, neoliberal development policy has arguably aggravated this challenge as public investments in rural infrastructure is sometimes shifting towards increased urban focus under the garb of resilience building for "economic growth centres" of metropolis. Although, with presently practiced concept of sustainable development allows a valuable global dialogue between north and south, it has not yet been able to resolve the historical gulch between the two by carving out a new meaningful equity led partnership within the development objectives of Sustainable Development Goals 2030 (Hayward & Roy, 2019). This gets reflected in winners and losers dilemma amongst small-holder farmers, gendered bias and entrepreneurs in rural and peri-urban areas for choosing climate-smart production systems and have are raising questions of sustainable partnership. Fast growing developing countries (Roy et al., 2021) have now become more focused on distributive justice (Roy & Foreman, 2021) implementation of prototypes in ground-zero, demanding that the effectiveness may not only be measured in terms of resilience to economic and environmental variables but also in terms of the inclusive growth leaving no one behind. Thus, societal development, equality as well as equity principles need stronger deep rooted embedding within the concept of sustainable development.

One experimental example, which needs to be further time tested for long term sustenance and scale up, can be drawn from the promotion of low emission, low chemically protected intensive organic farming. Farmers are trying to scale up sustainably intensifying the agricultural productivity with increasing assurance of better yields, enabled by better market returns returns on investments. However, farmers chose to be in direct link with the market mechanism so that they have market power to influence the price mechanism. Additionally, they want value added services like microinsurance coverage, subsidized health care, better education and socio-political-economic recognition of their value addition to the global/local environmental/ecosystem service through alternative environment friendly organic farming practices. This demand is distributive justice driven as this ecosystem value addition and seeking compensation for the risk farmers are enduring in experimenting with organic farming in their own lands for global good. In current the regime of industrial inorganic farming experimentation on their own fields the organic products as well as to compete in the market adds burden of uncertainty of success and failure for global good falls unjustfully on the farmers who are leading the experimentation. The burden of risk and uncertianty is unjustifiably falling on the farming communities. Socio-political-economic reality has not yet been strengthened by social protection mechanisms to take care of distributive justice by way of covering the uncertainty of retrieving a more assured return on investment in parallel with procedural justice of showing inclusiveness in promoting sustainable practices.

We envisage and advocate for a new ecomodern economic paradigm that builds on the principle of distributive justice besides procedural justice and introduces a 'payment' mechanism for the 'opportunity cost' of going 'green' for the local community. This may be through market mechanism or through public expenditure or social protection mechanisms or a combination of all as the context demands. As we have discussed earlier, one of the approaches to compensate the loss is to ensure natural rights of commons by actualizing a governance mechanism involving communities with flexibility and dynamism built in. However, the entitlements of communities in governance of natural resources in rural ares for equitable agriculture and resource sharing is constrained by several power dynamics associated. Community's aspiration for improving collective wellbeing, amidst the mainstream individualised goal seeking economic development creates confusion and sometimes ends up in downscaling of traditional community's collective goal to individualised benefit seeking practice through unsustainable use of nature's goods and services. Imagining a new institutional arrangement that compensates a community for the opportunity cost of going green can enable these activities to grow while addressing the hesitancy to change.

5.2 Cost of Conservation and Commons

Whenever we talk about environmental conservation, it implies following ecosystem-service based approach over a bio-geographical area defined artificially by political negotiation or otherwise. Ecosystem service based approach relies on maintaining, protecting, restoring to allow natural regenerative capacity to stay functional to ensure functionality and sustainability through nondeclining stock of natural capital . Non declining natural capital principle allows the opportunity for equitable intergenerational access to productive base which in turn can generate a benefit flow that can be managed socially and economically within any political or governance system to share benefits equitably. It can be sensitive intervention point (Farmer et al., 2019) based approach or system wide approach depending on the natural system's characteristic. All activities pertaining to environmental conservation therefore, needs knowledge generation and sharing and application which do involve costs. Given the public good nature of the outcome public investment is the way forward involving overlapping generations as stakeholders.

However, at the margin, the cost of environmental protection is higher than indicated by direct measures. This is because there are both environmental and non-environmental components of costs in conservation interventions. In the long run investments in conservation activities to be sustainable will need to get the return that at least compensates for the total costs. Looking in terms of competitiveness or productivity of inestible funds the economic return will need to be higher than the opportunity costs to the investor (Ref: https://media.rff.org/documents/RFF-DP-98-36-REV.pdf).

Cost of conservation needs to be seen in a systemic framework. Costs of conservation invariably consists of a number of components, such a cost of restoration and conservation activities on field, cost of infrastructure development for conservation needs, costs for prevention and sustainability of remedial measures taken, which also includes the cost of monitoring and evaluation, administrative costs, societal costs and other contingencies (https://www.env.go.jp/en/policy/ssee/eag02.pdf). This in many a times do not purportedly include cost of compensating the opportunities lost by the inhabitants. Since this is an intangible cost, usually it is not booked in conservation budget and planning. On the contrary, this cost estimation is crucial in a way that if this is to be borne by the inhabitants and beneficiaries then the partnership and participation may be withdrawn rendering sustainability of the intervention in a dismay. Thus, it is a significant node in developmental economics to understand what is opportunity cost and how can it be internalized in conservation budget and planning.

5.3 Opportunity Cost: Concept, Assessment and Internalization

When resources are scare and facing competing demands, decision to allocate that resource in one activity deprives the society of getting return from another activity where resource is not allocated. This gives rise to the concept of opportunity foregone. Constraints of material, fiscal, monetary resources compel us to take decisions for resource allocation where the return is the highest. The cost of this forgone opportunity is termed in economic literature as economco cost/opportunity cost which is different than financial cost. Therefore, under scarcity for every chosen option, practically there will be some opportunity cost which is not always considered in financial accounting system. Thus, when we plan for conservation and choose environmentally befitting options or an option that enhances/helps in maintaining the natural capital to keep service flow non declining then also there may be some foregone opportunities in terms of non environmental benefits like economic benefits and thus has an opportunity cost. Similarly, when the marginal communities adopt a resilient practice or even willingly shift to a sustainable solution which may be not be providing enough economic return in the absence of appropriate market mechanism for valuing the return to natural capital there would be some incumbent market opportunities foregone. On the reverse also any environmentally unsustainable practice also has opportunity cost which are not valued correctly by incumbent market practices, governance structure, regulatory system. Therefore, to consider these issues in the planning and budgeting practices with proper valuation of the opportunity cost need to be done with appropriate accounting system which goes beyond financial accounting, market price based values by consideration of valuation of opportunity costs in terms of environmental or natural service/product foregone through true economic cost estimation (Hanley et al., 1999; Dasgupta et al., 2015). Since communities at present are still functioning within a market-regulatory-governance system where environmental flow is either under valued or not valued at all in the incumbent system the community generating environmental services faces less return and experience trade-offs as they bear higher opportunity costs in terms of economic benefits foregone. The social cost of any foregone opportunity depends on its value and market price both and may diverge if value and prices are not the identical, which is generally the case. Value may vary with subject whose choice is represented to represent the worth. It can be choices of individual, societal or any other intangible factor dependent, whereas the price indicates its market exchange rate. Quite often the value can also vary depending on what is considered as constraint: money or time? What is foregone? Money or time in availing an opportunity. A rough estimate can be for example, if the delay in arrival of a train is by 40 minutes and 700 passengers get stranded for this then the price of time would be equal to the product of number of passengers, waiting time and their average dues (say salary) for that period of waiting.

5.4 Opportunity Cost and Production Possibility Frontiers

It can be named in a variety of ways but we are using 'green opportunity cost'. It can arise when decision is to be taken for allocating a 'commons'/common property natural resource for producing say two products/types of outputs and a combination is favoured over the other. As we are assuming that natural resource endowment is fixed, how to convert the resource into output that technology is also known and fixed, two alternative outputs can be produced in various combinations depending on societal need. However, at any point of time only one combination is possible. The opportunity cost of engaging natural resource and making a choice in favour of one combination of outputs (e.g., 10 units of agricultural produce and 20 units of managed forest land) over another combination (e.g., just the reverse 20 units of agricultural produce and 10 units of managed forest land) would imply a decision to sacrifice some output over the other (here in the example sacrificing managed forest land to get more agricultural output). This green opportunity cost is illustrated using a production possibility frontier (PPFs). PPF is concave (Fig. 5.1) to the origin because as one increases the share of one output the relative sacrifice of the other goes up and the latter's scarcity value goes up which means opportunity cost of producing agricultural produce goes up as managed forest land is sacrificed more and more. Here in the above example managed forest land becomes scarce and agricultural produce become increasingly abundant and opportunity cost of agricultural produce goes up.

The gradient of the PPF gets steeper as more of Production Option 2 is preferred over Option 1, indicating a greater sacrifice in terms of opportunities foregone of production option 1. Opportunity cost can thus be thought of in terms of how

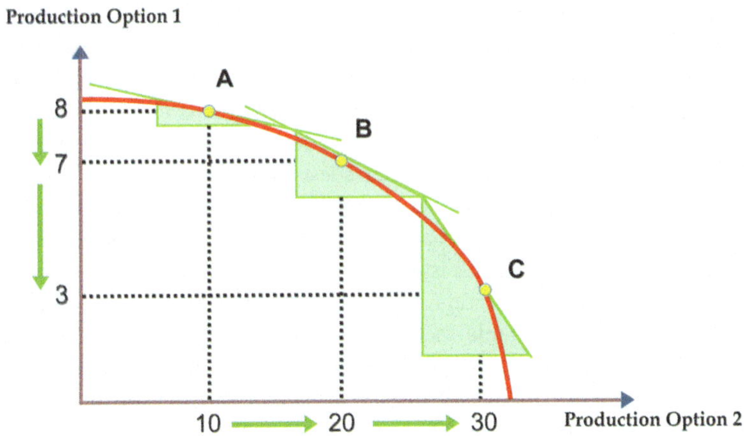

Fig. 5.1 PPF Curve & Pareto Efficiency of allocation

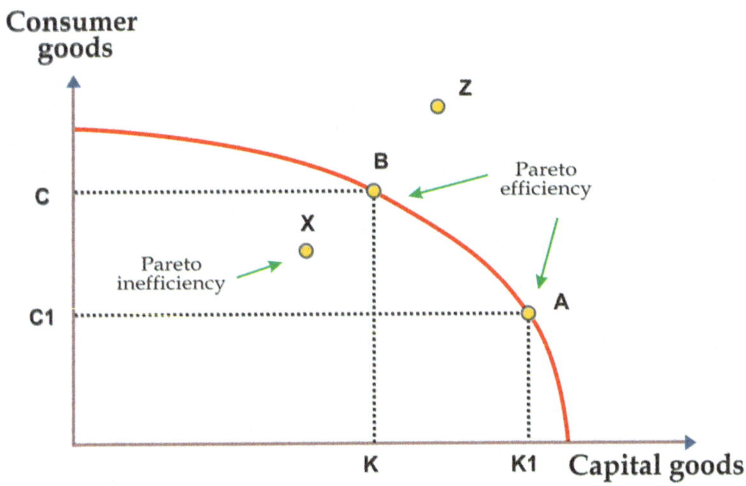

Fig: 5.2 PPF Curve & Pareto Efficiency

decisions to increase the production of an additional/marginal, unit of one good may lead to a decrease in the production of another good.

Points on the PPF, such as points 'A' and 'B', are said to be efficient (maximum possible output levels given the resource endowment and the technology) and indicates that scarce resources are being allocated to get the full potential out of the resources as optimum level of production is achieved. Leaving no room for further possibility of increasing the production going beyond the frontier unless resource endowment becomes bigger or technology becomes better and staying below the frontier is not a preferred option as it leaves wasted/underutilised/under employed resources. The points on the frontier are also called *Pareto efficient points* 'A' or 'B' (Fig. 5.2). Any point inside the PPF, such as point 'X' is said to be inefficient because the economic system can increase the output by just reallocating the economy's existing resources. Any point outside the PPF, such as point 'Z', is not achievable with the economy's current scarce resources, and technology although there may be higher aspirations. Thus, Pareto efficiency means when an economy is operating at its optimum i.e., on the frontier and no increase in output can be achieved by reallocation of the existing resources. PPF help in delineating the feasible region, optimum, and the impossible regions with the resource endowments with available technology to convert them in output.

5.5 Economic Values of Nature's Goods and Services

The monetary value of flow of all the service from natural capital and ecosytem-services have so far not been recognized by the macro accounting system globally given that it follows a system of national accounting system represented by a metric GDP (gross domestic product) which represents aggregate market value of goods and services produced or consumed within an economy. So current practice is narrowly focused on valuation method which functions within a market institution. Mostly the costs and benefits of nature conservation like species rehabilitation or habitat restoration are not visible/exchanged in the market (may not be possible or desirable also to have property ownership defined market institutional arrangement for all goods and services). Partially, they captured by market led tourism activities and revenues. That is why underestimation or value of services from nature is inevitable unless special care is taken. Hence cost-effectiveness solutions in practice is most likely will undervalue conservation and sustainable solutions of the local people thereby making them unattractive propositions.

5.6 Valuating Opportunity Cost for PES

Arriving at estimate or accounting for the fair and equitable compensation for those who are providing the ecosystem services in the absence of market, incomplete market, approrpiately designed regulatory, governance mechanism is complex and need new techniques and methods. Also, valuing contribution through the conservation efforts also get complicated due to the variable land costs, dealignments among multiple stakeholders, methodological constraints and inequitable sharing of outcomes (Corbera et al., 2007; Adams et al., 2010; Dong et al., 2011; Bryan, 2013; FFI, 2014). So, the estimates of the opportunity cost at the margin that is borne by the beneficiaries in course of sustainably enhancing the ecosystem services are left at lurch, though its estimation is crucially important for the economic viability of design of market-based incentive instrument for PES and biorights scheme (Pagiola & Bosquet, 2009; Plumb et al., 2012).

Attempts at calculating the opportunity cost and related other estimates like cost of conservation, burden on trade-offs etc. have so far included returns on investment on farming practices, proxies of land attributes, land price etc. along with screening contracts and procurement auctions, as well (Chomitz et al., 2005; Naidoo & Adamowicz, 2006; Jack et al., 2009; Wünscher et al., 2011; Borrego & Skutsch, 2014). The main constraints in standardizing a method on these, lay with detailed information, extensive data inputs, large data pool etc. Strengthening of the valuation model depends on the accuracy of data. Few model/methods are discussed here.

1. Flow-Approach: This is based on net return from land-use and uses an array of assumptions, for which extensive data input is necessary. It must account for short-term variability of returns which would need comprehensive local knowledge (Plumb et al., 2012).
2. Rent based Approach: This is based on sale or hire values from existing and active land-markets. The challenge lies in correlating it with the opportunity cost of conserving the land or area with a precise and reliable method that is neither expensive nor extensive (Wünscher et al., 2011).
3. Combination approach: This is simply aggregating both to get a more robust valuation method. In case the valuation is being done for PES scheme then provider's willingness to accept a fee for conserving as PES is incorporated in the method.

However, the above methods have also been debated for the risk of generalization as the opportunity cost would invariably vary spatially with changing socio-demographic facets and as well vary with time, market, non market institutional arrangement. So, the need for further research and innovation in finding a practical robust and methodological improvement for evaluating opportunity cost, still remains. As a part to this end, it is essential to quantify the impact on beneficiaries of the intervention and the ecosystem attributes for developing an incentive scheme for compensating true opportunity costs, especially where robust and active land markets are absent/distorted. This is because opportunity costs in provider groups, and the ecosystem services provided, are often heterogeneous (Tallis & Polasky, 2009; Adams et al., 2010; Chen et al., 2010).

5.7 Paying for Biorights from Ecosystem Services

Whenever, there is a payment or compensation mechanism to be designed for biorights, a principle need to be worked out in a place or context specific based perspective and needs to be understood how generalizable they might be. In principle, these payments can be to compensate for lost opportunities of the local communities in pursuing with conservation objectives, while allocating time and effort to safeguard natures goods and services. Else, this can also be total value: both direct and indirect- created by safeguarding the natural capital. While the former is context dependent latter has more global or macro implication. If value created is in the form of global good then the payments for biorights may come from global communities for conservation of global common goods say e.g., decarbonization through enhanced global carbon sink service creation (or more generally GHG emissions). If it is for local environmental protection or local biodiversity enhancement then it may also come from extended downstream users as payment for ecosystem services by the beneficiaries to the providers. At this point, it must be noted that paying for the cost of conservation or compensating the lost opportunities of the communities is in no-way a license to unsustainably use the natures goods and services. Payments for ecosystem services or compensation for the lost opportunity costs are for the local communities to support conservation, who are otherwise underprovided of resources for conserving the goods and services of nature. Let us discuss the payment logic based on the assumptions of Eijk& Kumar (van Eijk & Kumar, 2009) in the following figure Fig. 5.3 below.

Suppose, we consider a perennial wetland in a riverine floodplain that sustains a local agrarian community with its provisioning services like food, fish, fodder production etc. and are thus conserved by the community. The wetland has extended regulatory services at meso scale acting as flood mitigation mechanism. It can also provide supporting services like supply source of water. The total benefit of this

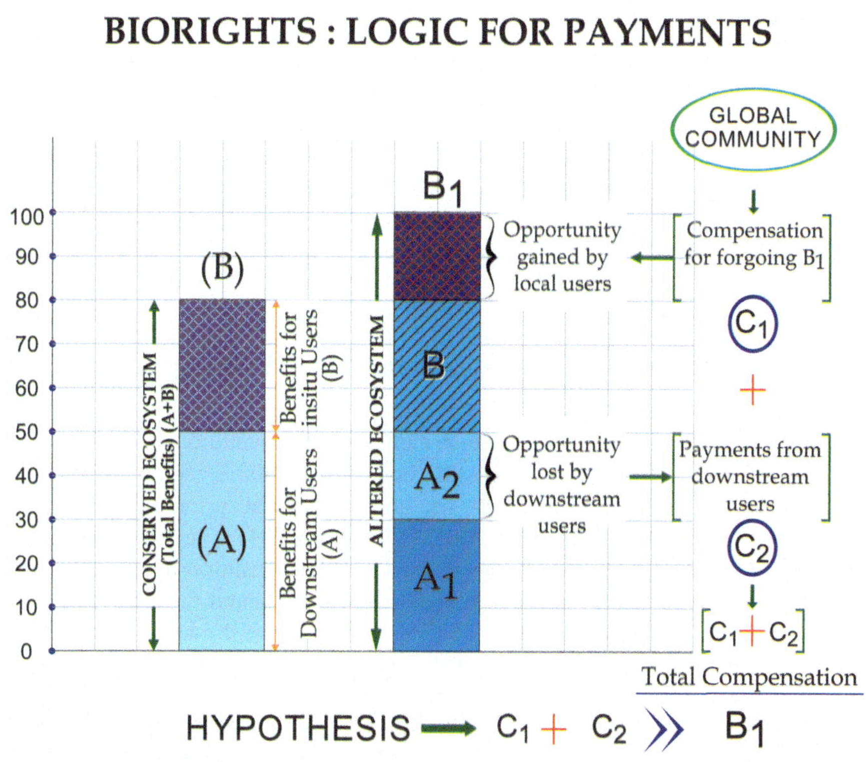

Fig. 5.3 Logic for Payment of Biorights

conserved flood plain (a natural capital) includes benefits of downstream users [A], and in-situ benefits for the local users [B]. Now, let's assume that the local community [B] plans to change the ecosystem attributes of the natural capital to get more benefits through water storage, captive fisheries etc. in situ and decides to alter the land-use and management strategies in the flood plain. This would obviously have an impact on the total benefits as suggested before as service flow to user group A might be reduced. Let's consider that in the altered ecosystem, downstream benefits are now reduced to A1 and the downstream user loses opportunities of using ecosystem services by A2, whereas, the local agrarian community will be additionally benefited by gaining the opportunities created in the altered ecosystem, say B1. Now, if the conserved ecosystem needs to be kept unaltered, the opportunities lost by the community for not altering the ecosystem, that is B1 needs to be equitably compensated and as well it must also motivate them to continue their conservation efforts so that the downstream users too do not get deprived of their opportunities that is A2. In this case, to regain this opportunity A2, downstream users may consider paying the local communities for their foregone opportunity B1, so that they continue providing the ecosystem services through conservation initiatives. Now, if C1 is the recompense from global communities for loosing opportunities as in B1 and C2 be the payments from downstream users for buying the services from the provider, then the value of both C1 and C2 must be more than B1 to motivate the local communities in conservation.

The challenge comes in calculating the payments for compensations, namely C1 and C2. While C1 can be calculated by estimating 'willing to pay' value of the global communities C2 can be calculated based on market prices of the opportunity lost by the downstream users. In this regard, methods for weighing ecosystem services in the local context and deriving a referral value over space and time become essential. Though this is meagrely addressed in environmental economic research, but some approaches to deal with it are explained in the following discussions.

5.8 Incentivizing Ecosystem Services: A Critical Appraisal

Intensive agricultural practices as an aftermath of green revolution has ravished the agro-ecosystem to an extent that its restoration to save natures capital and services through land use management has become the global priority (Millennium Ecosystem Assessment, Reid et al., 2005; Foley et al., 2011; Ehrlich et al., 2012). Failure to internalize the environmental costs in the agricultural market economy pertaining to conservation, restoration and wise use of natural resources has been responsible for such degradation (Lant et al., 2008). Several market-based policy instruments for incentivizing these costs have emerged to prevent such market failure (Farley & Costanza, 2010). While these are novel approaches and raises hope (Daily et al., 2009; Pascual & Perrings, 2007), but potential negative impacts of incentivizing community efforts and apprehended inefficacies of these policy instruments can't be ruled out (Frame, 2011; Kinzig et al., 2011).

Incentives motivate land use management changes in agro-ecosystems in a complex way, thereby triggering multiple such changes, both intended and unintended, that has potential benefits and co-benefits, as well trade-offs and synergies across multiple ecosystem services (May & Spears, 2011). Coupling effects of multiple incentives have been found to impact land use management changes (Deal et al., 2012) compounding a unique matrix between incentives and ecosystem services, which are usually nonlinear and may show spatio-temporal variability (Holland et al., 2011; Laterra et al., 2012). To abate the negative impacts and as well assure substantial gains from these incentives, the design of the financial instrument needs to have a dynamic feedback control mechanism (White et al., 2012; Bryan & Crossman, 2013).

Market prices for public goods like biodiversity and common pool resources like fisheries seldom rise and as farmers do not see an attractive demand for these non-market ecosystem services, they downscale the production of these commodities (Ribaudo et al., 2010). Market instruments then incentivize the production motivations to avoid market failure (Farley & Costanza, 2010). Only meticulously designed regulatory framework can have control mechanisms in place for preventing market failures and needs to incorporate safe minimum standards, quantifiable units of service provision, clearly defined property rights, monitoring requirements, and contractual arrangements thereto (Kroeger & Casey, 2007). Financial market instruments for incentives those are properly supported by regulation can only be used to motivate the production of ecosystem services beyond critical levels (Farley, 2008).

In any given region, multiple incentives – from global commodity markets to locally implemented incentives for public good and common pool resources – may co-exist for governing the production of ecosystem services. However, there are instances of politicizing the demand of a particular ecosystem service either for commercial benefits or vested social interests so that the price signal is raised in the market. As a ripple effect, the design of the financial instrument gets adversely impacted. In the case of East Kolkata Wetlands, when ecotourism was promoted as an alternative economic opportunity for fishers to recompense the opportunity lost in protecting the wetlands in the lean winter season, the same was politicized as a matter of common benefit and all fisher's cooperatives were engaged in wetland ecotourism. This perversive tendency was found to be led by the vested interest of changing the land-use pattern of the wetlands in the pretext of ecotourism (Amerasinghe & Dey, 2018).

In agro-ecosystem land use and management activities there can be both winning and losing outcomes (Tallis et al., 2008), as well negative correlations in ecosystem service changes may also occur (Raudsepp-Hearne et al., 2010), if the land use change increases the efficacy of one or one set of ecosystem services at the expense of another then there will be trade-offs. Such trade-offs and synergies vary largely over multiple spatio-temporal scales (Power, 2010; Rodríguez et al., 2006). Such trade-offs from agro-ecological interventions have been have been reported as replacement of many supporting and regulatory services and even cultural services in course of production of energy crops, fodder and fibre etc. (Bennett & Balvanera, 2007; Lant et al., 2008). More site specific trade-offs like failures in sediment trapping or local species disappearance have also been reported from agro-ecosystems (Swallow et al., 2009; Barraquand & Martinet, 2011). Similarly restoring natural ecosystems from agro-ecological intervention area too has led to trade-offs like fall in agro-biodiversity, salinity intrusion, change in carbon sequestration potentials etc. (Maron & Cockfield, 2008; Crossman et al., 2011; Nelson et al., 2008). If the net gains of land use change are maximized and multiple ecosystem services are considered, more efficient outcomes can be achieved (Chen et al., 2010; Crossman & Bryan, 2009; Nelson et al., 2008; Wainger et al., 2010).

Individual policy instruments designed for incentivizing any one ecosystem service can seldom achieve multiple objectives efficiently (Nelson et al., 2008). If implied, they will have unintended consequences compromising with their primary objective, turning the incentive into a bad investment. This is only because, intra and interlinkages of ecosystem services and incentives are very complexly intertwined making it acutely specific. Owing to the unintended consequences broad impacts are missed out in the outcome, rendering it unfavourable for the community at large and the net social value produced across multiple services too, decreased. Provisions of bundled payments for multiple ecosystem services in the incentive design can create price premiums for sellers but can eventually increase the transaction costs towards rigorous monitoring and supervision (Kemkes et al., 2010; Farley & Costanza, 2010).

5.9 Biorights Model in Economic Environmental Zones and Poverty Alleviation

Nature as a resource has varied uses. It provides us all our basic necessities. It has always been the primal base for all our creations. But today the tables have turned. Our mindless exploitation of environment makes this base unstable. This exploitation is not just limited to natural resources. We experience unequal income distribution in society and experience poverty. The intense combination of environmental degradation and poverty could not have been more conspicuous than at present in the twenty-first century. Yet, as positive members of human race we can't help but hope and faith in the creative abilities of human beings, as a reflection of humanity's age old struggle for existence. With this deep sense of hope we try to present a neo-classical economic model on poverty alleviation in Special Environmental Zones. The class of economic agents who will be our protagonist belongs to Below Poverty Level (BPL). Moreover they reside in areas of Global Environmental Importance namely the Environmental Zones (EZ). However in our model we are interested in members of BPL in **Economic Environmental Zones (EEZ)**. The subtle difference being that in EEZ the environment has a dormant marketable economic value that is yet to be utilized. EEZ $\subseteq$ EZ. This is a reasonable restriction as it would be absurd to accept every area under EZ to have marketable economic value. For example a location in the middle of a desert possibly won't have any marketable economic value. The line of demarcation between EEZ and EZ is assumed to be fixed.

Another important tool of our model is **Activation Technology (A)**. This crucial factor is like a key! It opens up the dormant economic value in areas under EEZ. The point is this technology helps us identify environmental markets and nature resources having marketable value. Now, a given amount of 'A' at a specific time point unlocks certain amounts of areas (ψ) under EEZ. Activation Technology activates dormant areas under EEZ. BPL members in such active areas can provide their labor input in the newly generated Nature Sector. The model tries to prove that wage earned by BPL members through labor in Nature Sectors helps them grow out of poverty. We try to compare between wages of Nature Sector (Z_E) and Normal Sector (Z_N). Here Z_N represents the existing market economy. We will find that Z_E gives more return to BPL members than Z_N. Since real factor returns are equal to their marginal product thus the earlier statement implies BPL members are more productive in Z_E. This is a sensible consideration as members of BPL are locals of that area. They know their environment better than anyone else. For example consider BPL members in a forest, let Z_E be eco-tourism in that forest. It is sensible to consider residents of the forest to know their forest better than others.

5.9.1 The Model: Assumptions and Core Concepts

A. **Economic Environmental Zone (EEZ):** It is a collection of Ecoregions* (ψ_α) of Global Environmental Importance that has dormant marketable value. $\psi_\alpha \in EEZ.\psi_\alpha$ is homogeneous in all respect barring one. Different ψ_α has different degree of dormancy (α) in marketable value of its environment. Let every point on a unit length be α. Thus ψ_α is atomistic and EEZ has infinite ψ_α's in it. Now, α close to 0 implies ψ_α has lower degree of dormancy, whereas α close to 1 implies ψ_α has higher degree of dormancy. This is shown in Fig. 5.4
B. **Activation Technology (A):** It is like a key that can activate dormant Ecoregions (Ecoregions is the short form of ecological regions) Now activation technologyis a function of time, t.

Mathematically,

$$A = A(t)$$

We assume activation technology to grow at a diminishing rate.

i.e.,

$$\frac{dA(t)}{dt} > 0; \frac{d^2A(t)}{dt^2} < 0; \text{Also, as } t \to \infty,\ A(t) \to \bar{A},\ \bar{A} \in [0, 1]$$

Fig. 5.4 Intensity of dormancy increases

$$\forall \alpha \in [0, 1]$$

We assume the economy has A_0 amount of initial activation technology at time t_0. Without loss of generality, lets assume A_0 corresponds to $\alpha = 0$, level of dormancy. This implies no ψ_α in EEZ have activated Nature Sector (Z_E) at t_0. Thus each ψ_α only has an active Normal Sector (Z_N) which is assumed to represent the standard market economy. The activation technologywe are concerned with is not the technological innovation considered in standard markets. Our 'A' is a special type of technology that establishes market value of environment and brings such ecological regions under the reach of competitive markets. For issues of simplicity we don't consider standard technological innovation of Normal and Nature Sector to grow over time. Output in both sectors depends on usual factors of production but not on standard technology. Dormancy is a negative function of activation technology. That is as activation technology increases dormancy falls. In other words, we can also say activation of Nature Sector is a positive function of activation technology. As activation technology increases over time dormant Ecoregions towards the higher end of unit length gets their Nature Sector activated. Let's assume linear and equi-proportional relation between Activation Technology and Dormancy. i.e. for $\alpha = \alpha^*$, $A = \alpha^*$ amount of activation technology required to activate Nature Sector of ψ_{α^*}. This assumption simplifies our case. Also we can normalize activation technology to the scale of unit length. i.e.

$$A(t) = \alpha, \; \forall \alpha \in [0, 1]$$

Diagrammatically (Fig. 5.5),

Explanation: At $t = 0$, $A = A_0 = 0$ (on the basis of our previous assumptions). Now, as t increases A(t) increases at a diminishing rate. As $t \to \infty$, $A(t) \to \bar{A}$. At any given value of t we have an unique A(t). Since $A(t) = \alpha$, thus we also have an unique α. Since $A(t) \leq \bar{A}$, thus $\alpha \leq \alpha$. This means that even if we consider infinite time horizon the technological innovation cannot go beyond $\bar{A}$ and thus activation of dormant Ecoregions is possible up to α level of dormancy. Thus the whole of EEZ cannot be activated. Only a part of it can be activated in infinite time. Let this part be called EEZ*.

Now,

$$EEZ^* = \left\{ \psi_\alpha^* | \alpha \leq \alpha \text{ where } \alpha \in [0, 1] \text{ and } \alpha = \bar{A} \right\}$$

EEZ* is the ultimate amount of environment that can be brought under market mechanism given the constraint of human beings ability to innovate. Each Ecoregion, ψ_α*, that belong to EEZ* will get activated at some point of time. All active

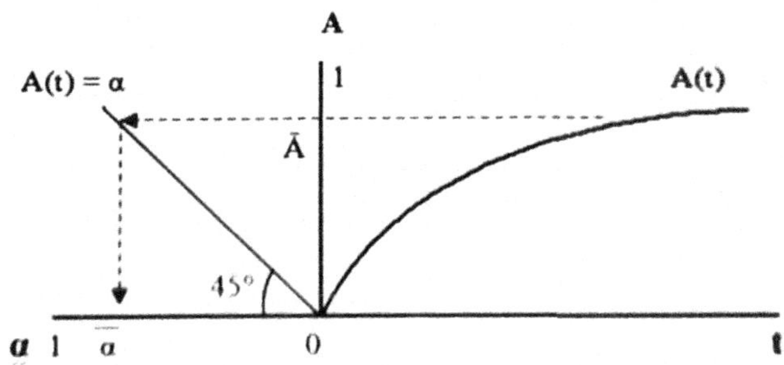

Fig. 5.5 Relationship between A(t) and α

Ecoregions are homogeneous. All dormant Ecoregions are also homogeneous. ψ_α* with lower α values are activated earlier, whereas ψ_α* with higher α values are activated latter in time. All dormant ψ_α* have an active Normal Sector, only. All active ψ_α* have both Normal and Nature sectors activated. Since, $A(t) = \alpha$, thus $t = A^{-1}(\alpha)$ [as, $A(\cdot)$ is invertible]. Thus ψ_α* activates at $t = A^{-1}(\alpha)$. Any ψ_α that don't belong to EEZ* never gets activated. This model allows poverty alleviation of BPL members in active ψ_α*.

C. **Members of BPL (L_{BPL}):** We assume the dormant ψ_α* have only BPL members (L_{BPL}) residing there. People above poverty line don't live in ψ_α*. They live in cities or other core economics zones. Thus dormant ψ_α* is an economically backward but environmentally forward region. Also we assume L_{BPL} knows their environment better than others.
D. **The Normal Sector (Z_N):** It is the always active economic sector. This sector is outside ψ_α*.Let the representative output produced in this sector be Y_N. Labor from ψ_α* migrate each day outside ψ_α* to work in Z_N and comes back at night. Other than L_{BPL} the production of Y_N uses Normal Capital (K) and Normal Labor (L_N). The factor market is competitive, thus individual factor of production are factor price takers. L_{BPL} comprise an infinitesimally small portion of the labor requirement for Z_N, since it is a huge sector. Figure 5.6 depicts this situation.
E. **The Nature Sector (Z_E):** In ψ_α*, the Z_E activates at $t = A^{-1}(\alpha)$. This sector is inside ψ_α*, thus small. Let the output produced in this sector be Y_E. $Y_E$is a commodity that primarily depend on environment. Production process of Y_E does not harm environment, rather it upgrades nature. We consider Optimum Nature Utilisation (ONU) which means optimally using nature such that it do not degenerate. For example, Y_E can be eco-tourism in forests, fish supplied by fishing industry of wetlands, etc. L_{BPL} has an edge when they work in Z_E. This edge comes from their deeper knowledge of their environment. Factor used for production of Y_E other than L_{BPL} are Environmental Capital (K_α), Normal Capital (K) and Normal Labor (L_N). Here also, factor market is competitive. L_{BPL} is employed in both Z_N and Z_E. Z_E might not absorb whole of L_{BPL} population. L_{BPL} that don't get job in Z_E, work in Z_N. Figure 5.7 depicts this situation.

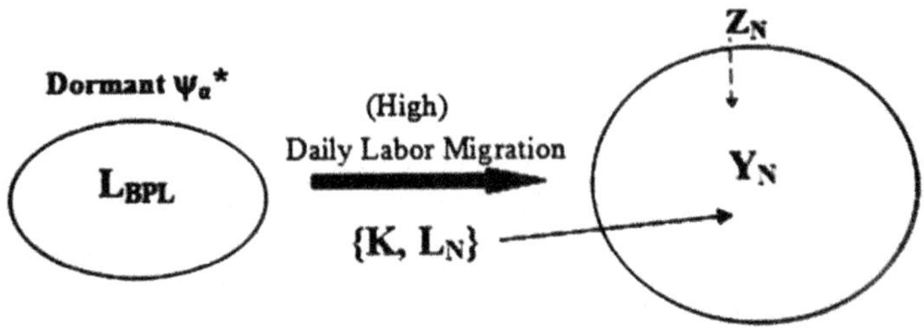

Fig. 5.6 Daily Labor Migration from dormant ψ_α* to Z_N

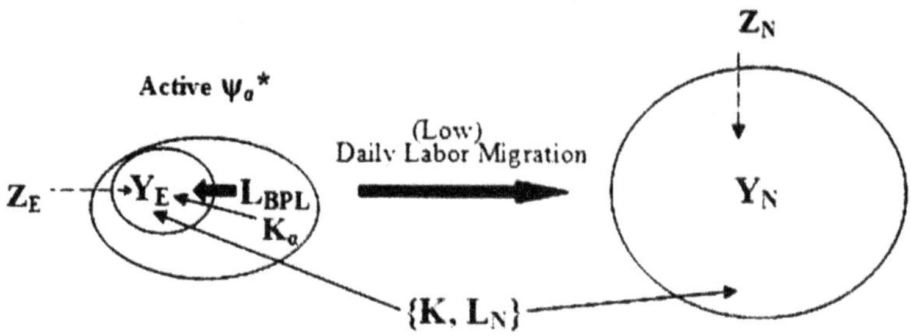

Fig. 5.7 L_{BPL} working in Z_E stays in active ψ_α*

Proposition 1
Under the setup of this model BPL members of ψ_α* will remain poor till $t < A^{-1}(\alpha)$.

Proof
Since $t < A^{-1}(\alpha)$, activation technology haven't reached the level that can activate Ecoregion, ψ_α*. Thus $\psi_\alpha^* \in EEZ^*$, is dormant. So, population of ψ_α* must work in the active Normal Sector. This is the only sector that is active at present. BPL population of all Eco-regions in EEZ work in Normal Sector. The Eco-regions are activated with time. Therefore, some members of BPL shift from Normal Sector to their respective Nature Sector. But this movement of BPL labor out of Normal Sector doesn't affect their wage rate. This is so since the Normal Sector is huge and labor from BPL make a very small portion of its total labor force. Thus Normal Sector is insensitive to change the BPL member's employment level. Thus we can abstract away from the changing dynamics of BPL employment.

$$\text{Production Function of } Y_N : Y_N = Y_N\left(K, L_N, L_{BPL}\right)$$

$$\text{Cost Function of } Y_N : C_N = rK + w_N L_N + w_\eta L_{BPL}$$

Where,

r = Rental rate of Normal Capital,
w_N = Wage rate of Normal Labor, and
w_η = Wage rate of BPL members residing in ψ_α*.

As total population of the economy is constant over time, therefore demand for Y_N is constant. Thus a steady market clearing price of Y_N is maintained over time. Let us call it P_η.

$$\text{Profit Function of } Y_N: \qquad \Pi_N = P_\eta{}^* Y_N - C_N$$

$$\text{or, } \Pi_N = P_\eta{}^* Y_N\left(K, L_N, L_{BPL}\right) - rK - w_N L_N - w_\eta L_{BPL}$$

At equilibrium, FOC = 0

$$\text{or, } \frac{\delta \Pi_N}{\delta L_{BPL}} = 0$$

$$\text{or, } w_\eta / P_\eta = \frac{\delta\left\{Y_N\left(K, L_N, L_{BPL}\right)\right\}}{\delta L_{BPL}} = MP_{BPL\eta}$$

Here, real wage received by L_{BPL} is equal to the marginal product of BPL members. People in BPL don't have apt capability, skill and knowledge. They are less productive in normal sector. Thus, they add very little to normal economic activities. Marginal product of L_{BPL} in Normal Sector is low. Therefore, w_η keeps them inside the vicious cycle of poverty.

For example, labors from East Kolkata Wetlands regularly migrate to Salt Lake City (a local township) for work. They work as servants, in tea-stalls or other such places where they earn very low real wages. Thus remains poor.

Proposition 2
When $t \geq A^{-1}(\alpha)$, Nature Sector of ψ_α^* gets activated. Thus, the poverty alleviation process for residents of ψ_α^* also gets activated. And in due course of time poverty is alleviated from ψ_α^*.

Proof
At $t \geq A^{-1}(\alpha)$, the economy achieves the technology required to activate ψ_α^*. Thus ψ_α^* now has its Nature Sector activated. Given the emergence of a new sector inside the eco-region, its resident BPL members have another form of employment. Since the Nature Sector is eco-region specific thus it's very small. Let the output produced in Z_E be Y_E. Nature sector is competitive but also highly dependent on environment. It performs optimum environment utilisation to achieve Y_E. Moreover Nature Sector has a welfare aspect in it initially. It is common knowledge that local BPL members have an inner understanding about their environment. Because of their poverty they are not well trained to instantaneous capitalization on the new nature sector. Thus, Nature Sector initially tries to train the local L_{BPL}. The sector bears this cost as it knows in near future the higher marginal product of BPL members will compensate for the initial training cost. BPL members will have higher marginal product in Nature Sector since here they will express their internal skills. For example, if the Nature sector is eco-tourism in forest eco-region then BPL members who live in that

forest can work as guides and managers as they know the forest better than all other. Since the Nature Sector is small thus there might be the case that the whole of L_{BPL} is not absorbed. Then those left out can work in the Normal Sector.

$$\text{Production Function of } Y_E: \quad Y_E = Y_E\left(K, K_\alpha, L_N, L_{BPL}\right)$$

$$\text{Cost Function of } Y_E: \quad C_E = rK + r_\alpha K_\alpha + w_N L_N + w_\varepsilon L_{BPL}$$

Where,

r_α = Rental rate of Environmental Capital. It can be zero if environment is everyone's property or some positive number if environment belongs to government or state.

w_ε = Wage rate of BPL members residing in ψ_α* and working in Z_E.

It is assumed price of Y_E is equal to price of Y_N, i.e. P_η.

$$\text{Profit Function of } Y_E: \quad \Pi_E = P_\eta{}^* Y_E - C_E$$

$$\text{or,} \quad \Pi_E = P_\eta{}^* Y_E\left(K, K_\alpha, L_N, L_{BPL}\right) - rK - r_\alpha K_\alpha - w_N L_N - w_\varepsilon L_{BPL}$$

At equilibrium, FOC = 0

$$\text{or,} \quad \frac{\delta \Pi_E}{\delta L_{BPL}} = 0$$

$$\text{or,} \quad w_\varepsilon / P_\eta = \frac{\delta\left\{Y_E\left(K, K_\alpha, L_N, L_{BPL}\right)\right\}}{\delta L_{BPL}} = MP_{BPL\varepsilon}$$

Thus, BPL members in Nature sector gets real wage equal to their marginal product in Nature Sector. Now, $MP_{BPL\varepsilon} > MP_{BPL\eta}$.Thus, $w_\varepsilon > w_\eta$. We know, w_η keeps BPL members in the vicious cycle of poverty, but w_ε help the BPL members to overcome poverty. We know in active eco-regions because of small size of Nature Sector all the resident BPL members might not be employed. Let γ fraction of L_{BPL} be employed in Z_E. Then $(1 - \gamma)$ fraction will be employed in Z_N.

$$\text{Community Income}\left(\text{in per BPL labor terms}\right) = \gamma\, w_\varepsilon + \left(1-\gamma\right) w_\eta, \text{ where } \gamma \in \left(0,1\right)$$

It is reasonable to consider that Activation Technology will term a Nature Sector to be active only when employment generated by the Nature Sector leads per capita community income to be above income corresponding to poverty line.

Our Model tried to show how activated nature sector in an eco-region alleviate poverty of members of BPL residing in that eco-region. We showed poverty alleviation is possible in a very simple and abstract framework. But our model can be made more complex and practical. We will state the possible variations of the model that can be considered.

- We assumed both L_{BPL} and L_N to remain constant over time. But L_{BPL} and L_N can grow at constant but different rates. Generally L_{BPL} grows at a higher rate than L_N.
- We assumed away from labour augmenting technological progress in the production functions. Yet we can introduce such technology. Also it would be sensible to assume labor augmenting technology for Normal Sector will have higher growth rate than that of Nature Sector.
- We could also have considered a population of BPL members residing outside eco-regions such as in core economic zones like cities or towns. They can provide their labor in Normal Sector though not in Nature Sector.
- For Normal Sector of smaller sizes it can be assumed that as LBPL shift to Nature Sector their shortage might increase their wage rate in ZN. Thus changing dynamics of BPL employment can bring changes in BPL wages.
- The economic activities undertaken in Nature Sector can be nature services which individual BPL members can exchange in return of BIORIGHTS (a financial tool that values nature services).
- We considered active Nature Sector in an eco-region will employ resident BPL members of that eco-region. But case where BPL members of other eco-regions can opt for employment in that active Nature Sector is also another interesting exercise.

Many other possible variations certainly exist. We tried to introduce the concept of Activation Technology and poverty alleviation through BPL service in active Nature Sectors. We felt eco-regions possess inner economic potential that when used optimally can help us alleviate local poverty. The economic potential is such defined as one that cause environmental upgradation. Thus we achieve the dual task of poverty alleviation and environmental upgradation through this model.

Empirical Validation Throughout history man has tried and tamed nature. Though mostly he has exploited environment harmfully yet there are cases of economic utilisation and sustainable environment development. We mention a few random cases here.

- Travel and Tourism industry at Sundarbans Reserved Forrest.
- Water reservoir and sprinkling irrigation system, in 1995, at Haji Noor Mohammed village in Khinjar Lake helped agricultural production.
- In November 2002, the Ramsar Wetland Conservation Award was presented to the Chilika Development Authority for "outstanding achievements in the field of restoration and wise use of wetlands and effective participation of local communities in these activities".
- East Kolkata Wetlands have effectively reused both the solid and liquid waste of the metropolis, as a nutrient rich medium for agriculture and aquaculture.

Policy Suggestions From the point of view of our model the most important factor towards poverty alleviation and environmental upgradation is Activation Technology.

It is this Activation technology that activates dormant eco-regions. It starts the Nature Sector in an eco-region. We know an active nature sector increase income of local BPL members and pull them out of poverty. Thus it is the responsibility of the policy maker to achieve faster growth in Activation Technology. This can be done by investing in Research & Development that helps at identifying potential eco-regions to be activated. Government or NGOs can see through the initializing process of the Nature Sectors. They can move forward and take steps to see sustainable transition is achieved for BPL members as they shift to the Nature Sectors. This can be done through proper training, micro-finance, subsidies, minimum wage guarantee schemes, and other methods. The authority should see through that the output produced in Nature Sector receives proper marketing. Moreover Government should try to develop a taste for such eco-friendly commodities among the general mass through publicity mechanism.

Our model tried to provide an abstract foundation for pro-cyclical relation between poverty alleviation and environmental development. We introduced a new type of technology, i.e. Activation Technology, to achieve the task. It is evident if the findings of this model get implemented in real world then there is a hope of solving the twin problem. Environment is very important to us, to economize environment and yet maintain its poignant presence is what we wanted to achieve. We showed directing man to do what he does best is the best solution to any crisis. Here, our poor man from the ecological region knew less about everything but more about his own land. He knew his forest, his land, his lakes..we wanted to help him market what skills he has (i.e. work in the local Nature Sector) and not be part of the Normal Sector that's far different from his own unique birthplace. Allowing humans to remain specialized to what they are born with was the central theme of this model.

References

Adams, V. M., Pressey, R. L., & Naidoo, R. (2010). Opportunity costs: Who really pays for conservation? *Biological Conservation, 143*(2), 439–448. https://doi.org/10.1016/j.biocon.2009.11.011

Amerasinghe, P. H., & Dey, D. (2018). *Recommendations for the wise use of urban and peri-urban wetlands in Kolkata, India.* International Water Management Institute (IWMI). CGIAR Research Program on Water, Land and Ecosystems (WLE), p. 8 (WLE BRIEFING SERIES NO. 23). https://cgspace.cgiar.org/handle/10568/96987

Barraquand, F., & Martinet, V. (2011). Biological conservation in dynamic agricultural landscapes: Effectiveness of public policies and trade-offs with agricultural production. *Ecological Economics, 70*(5), 910–920. https://doi.org/10.1016/j.ecolecon.2010.12.019

Bennett, E. M., & Balvanera, P. (2007). The future of production systems in a globalized world. *Frontiers in Ecology and the Environment, 5*(4), 191–198.

Borrego, A., & Skutsch, M. (2014). Estimating the opportunity costs of activities that cause degradation in tropical dry forest: Implications for REDD+. *Ecological Economics, 101*, 1–9. https://doi.org/10.1016/j.ecolecon.2014.02.005

Bryan, B. A. (2013). Incentives, land use, and ecosystem services: Synthesizing complex linkages. *Environmental Science & Policy, 27*, 124–134. https://doi.org/10.1016/j.envsci.2012.12.010

Bryan, B. A., & Crossman, N. D. (2013). Impact of multiple interacting financial incentives on land use change and the supply of ecosystem services. *Ecosystem Services, 4*, 60–72. https://doi.org/10.1016/j.ecoser.2013.03.004

Chen, X., Lupi, F., Vina, A., He, G., & Liu, J. (2010). Using cost-effective targeting to enhance the efficiency of conservation investments in payments for ecosystem services. *Conservation Biology, 24*(6), 1469–1478. https://doi.org/10.1111/j.1523-1739.2010.01551.x

Chomitz, K. M., Alger, K., Thomas, T. S., Orlando, H., & Nova, P. V. (2005). Opportunity costs of conservation in a biodiversity hotspot: The case of southern Bahia. *Environment and Development Economics, 10*(3), 293–312. https://doi.org/10.1017/S1355770X05002081

Corbera, E., Brown, K., & Adger, W. N. (2007). The equity and legitimacy of markets for ecosystem services. *Development and Change, 38*(4), 587–613. https://doi.org/10.1111/j.1467-7660.2007.00425.x

Crossman, N. D., & Bryan, B. A. (2009). Identifying cost-effective hotspots for restoring natural capital and enhancing landscape multifunctionality. *Ecological Economics, 68*(3), 654–668. https://doi.org/10.1016/j.ecolecon.2008.05.003

Crossman, N. D., Bryan, B. A., & King, D. (2011). Contribution of site assessment toward prioritising investment in natural capital. *Environmental Modelling & Software, 26*(1), 30–37. https://doi.org/10.1016/j.envsoft.2010.04.022

Daily, G. C., Polasky, S., Goldstein, J., Kareiva, P. M., Mooney, H. A., Pejchar, L., … Shallenberger, R. (2009). Ecosystem services in decision making: Time to deliver. *Frontiers in Ecology and the Environment, 7*(1), 21–28. https://doi.org/10.1890/080025

Dasgupta, P., Duraiappah, A., Managi, S., Barbier, E., Collins, R., Fraumeni, B., Gundimeda, H., Liu, G., & Mumford, K. J. (2015). How to measure sustainable progress. *Science, 350*(6262), 748. https://doi.org/10.1126/science.350.6262.748

Deal, R. L., Cochran, B., & LaRocco, G. (2012). Bundling of ecosystem services to increase forestland value and enhance sustainable forest management. *Forest Policy and Economics, 17*, 69–76. https://doi.org/10.1016/j.forpol.2011.12.007

Dong, Z., Yan, Y., Duan, J., Fu, X., Zhou, Q., Huang, X., … Zhao, J. (2011). Computing payment for ecosystem services in watersheds: An analysis of the middle route project of south-to-north water diversion in China. *Journal of Environmental Sciences, 23*(12), 2005–2012. https://doi.org/10.1016/s1001-0742(10)60663-8

Ehrlich, P. R., Kareiva, P. M., & Daily, G. C. (2012). Securing natural capital and expanding equity to rescale civilization. *Nature, 486*(7401), 68–73. https://doi.org/10.1038/nature11157

Farley, J. (2008). The role of prices in conserving critical natural capital. *Conservation Biology, 22*(6), 1399–1408. https://doi.org/10.1111/j.1523-1739.2008.01090.x

Farley, J., & Costanza, R. (2010). Payments for ecosystem services: From local to global. *Ecological Economics, 69*(11), 2060–2068. https://doi.org/10.1016/j.ecolecon.2010.06.010

Farmer, J. D., Hepburn, C., Ives, M. C., Hale, T., Wetzer, T., Mealy, P., … Way, R. (2019). Sensitive intervention points in the post- carbon transition. *Science, 364*(6436), 132–134.

FFI (Fauna and Flora International). (2014). *Opportunity cost analysis: Lessons learned from REDD+ and other conservation strategies*. http://www.fauna-flora.org/wp-content/uploads/Opportunity-Cost-Analysis1.pdf

Foley, J. A., Ramankutty, N., Brauman, K. A., Cassidy, E. S., Gerber, J. S., Johnston, M., … Zaks, D. P. (2011). Solutions for a cultivated planet. *Nature, 478*(7369), 337–342. https://doi.org/10.1038/nature10452

Frame, D. J. (2011). The problems of markets: Science, norms and the commodification of carbon. *The Geographical Journal, 177*(2), 138–148. https://doi.org/10.1111/j.1475-4959.2010.00385.x

Hanley, N., Shogren, J., & White, B. (1999). *Environmental economics in theory and practice*. Macmillan Education.

Hayward, B., & Roy, J. (2019). Sustainable living: Bridging the north-south divide in lifestyles and consumption debates. *Annual Review of Environment and Resources, 44*, 157–151.

Holland, R. A., Eigenbrod, F., Armsworth, P. R., Anderson, B. J., Thomas, C. D., & Gaston, K. J. (2011). The influence of temporal variation on relationships between ecosystem services. *Biodiversity and Conservation, 20*(14), 3285–3294. https://doi.org/10.1007/s10531-011-0113-1
Jack, B. K., Leimona, B., & Ferraro, P. J. (2009). A revealed preference approach to estimating supply curves for ecosystem services: Use of auctions to set payments for soil erosion control in Indonesia. *Conservation Biology, 23*(2), 359–367. https://doi.org/10.1111/j.1523-1739.2008.01086.x
Kemkes, R. J., Farley, J., & Koliba, C. J. (2010). Determining when payments are an effective policy approach to ecosystem service provision. *Ecological Economics, 69*(11), 2069–2074. https://doi.org/10.1016/j.ecolecon.2009.11.032
Kinzig, A. P., Perrings, C., Chapin, F. S., Polasky, S., Smith, V. K., Tilman, D., & Turner, B. L. (2011). Paying for ecosystem services-promise and peril. *Science, 334*(6061), 1348. https://doi.org/10.1126/science.334.6061.1348
Kroeger, T., & Casey, F. (2007). An assessment of market-based approaches to providing ecosystem services on agricultural lands. *Ecological Economics, 64*(2), 321–332.
Lant, C. L., Ruhl, J. B., & Kraft, S. E. (2008). The tragedy of ecosystem services. *Bioscience, 58*(10), 969–974. https://doi.org/10.1641/B581010
Laterra, P., Orúe, M. E., & Booman, G. C. (2012). Spatial complexity and ecosystem services in rural landscapes. *Agriculture, Ecosystems & Environment, 154*, 56–67. https://doi.org/10.1016/j.agee.2011.05.013
Maron, M., & Cockfield, G. (2008). Managing trade-offs in landscape restoration and revegetation projects. *Ecological Applications, 18*(8), 2041–2049. https://doi.org/10.1890/07-1328.1
May, L., & Spears, B. M. (2011). Managing ecosystem services at loch Leven, Scotland, UK: Actions, impacts and unintended consequences. In L. May & B. M. Spears (Eds.), *Loch Leven: 40 years of scientific research. Developments in hydrobiology* (pp. 117–130). Springer. https://doi.org/10.1007/978-94-007-4333-5_11
Naidoo, R., & Adamowicz, W. L. (2006). Modeling opportunity costs of conservation in transitional landscapes. *Conservation Biology, 20*(2), 490–500. https://doi.org/10.1111/j.1523-1739.2006.00304.x
Nelson, E., Polasky, E. S., Lewis, D. J., Plantingall, A. J., Lonsdorf, E., White, D., Bael, D., & Lawler, J. J. (2008). Efficiency of incentives to jointly increase carbon sequestration and species conservation on a landscape. In *Proceedings of the National Academy of Sciences of the United States of America* (pp. 9471–9476).
Pagiola, S., & Bosquet, B. (2009). *Estimating the costs of REDD at the country level.* World Bank [online] http://www.forestcarbonpartnership.org/sites/forestcarbonpartnership.org/files/REDD-Econ_04-09-09.pdf
Pascual, U., & Perrings, C. (2007). Developing incentives and economic mechanisms for in situ biodiversity conservation in agricultural landscapes. *Agriculture, Ecosystems & Environment, 121*(3), 256–268. https://doi.org/10.1016/j.agee.2006.12.025
Plumb, S. T., Nielsen, E. A., & Kim, Y. S. (2012). Challenges of opportunity cost analysis in planning REDD+: A Honduran case study of social and cultural values associated with indigenous forest uses. *Forests, 3*(2), 244–264. https://doi.org/10.3390/f3020244
Power, A. G. (2010). Ecosystem services and agriculture: Tradeoffs and synergies. *Philosophical transactions of the royal society B: biological sciences, 365*(1554), 2959–2971. https://doi.org/10.1098/rstb.2010.0143
Raudsepp-Hearne, C., Peterson, G. D., & Bennett, E. M. (2010). Ecosystem service bundles for analyzingtradeoffs in diverse landscapes. *Proceedings of the National Academy of Sciences, 107*(11), 5242–5247. https://doi.org/10.1073/pnas.0907284107
Reid, W. V., Mooney, H. A., Cropper, A., Capistrano, D., Carpenter, S. R., Chopra, K., Dasgupta, P., Dietz, T., Duraiappah, A. K., Hassan, R., Kasperson, R., Leemans, R., May, R. M., McMichael, A. J., Pingali, P., Samper, C., Scholes, R., Watson, R. T., Zakri, A. H., … Zurek, M. B. (2005). *Ecosystems and human well-being – Synthesis: A report of the millennium ecosystem assessment.* Island Press.

Ribaudo, M., Greene, C., Hansen, L., & Hellerstein, D. (2010). Ecosystem services from agriculture: Steps for expanding markets. *Ecological Economics, 69*(11), 2085–2092. https://doi.org/10.1016/j.ecolecon.2010.02.004

Rodríguez, J. P., Beard, T. D., Jr., Bennett, E. M., Cumming, G. S., Cork, S. J., Agard, J., … Peterson, G. D. (2006). Trade-offs across space, time, and ecosystem services. *Ecology and Society, 11*(1) https://www.jstor.org/stable/26267786

Roy, J., & Foreman, C. (2021). *Ecomodern justice, breakthrough journal No.14*. Summer. https://thebreakthrough.org/journal/no-14-summer-2021

Roy, J., Das, N., Some, S., & Mahmud, H. (2021). Fast growing developing countries: Dilemma and way forward in a carbon constrained world. In P. Dasgupta et al. (Eds.), *Sustainable development insights from India: Selected essays in honour of Ramprasad Sengupta* (pp. 23–41). https://doi.org/10.1007/978-981-33-4830-1

Swallow, B. M., Sang, J. K., Nyabenge, M., Bundotich, D. K., Duraiappah, A. K., & Yatich, T. B. (2009). Tradeoffs, synergies and traps among ecosystem services in the Lake Victoria basin of East Africa. *Environmental Science & Policy, 12*(4), 504–519. https://doi.org/10.1016/j.envsci.2008.11.003

Tallis, H., & Polasky, S. (2009). Mapping and valuing ecosystem services as an approach for conservation and natural-resource management. *Annals of the New York Academy of Sciences, 1162*(1), 265–283. https://doi.org/10.1111/j.1749-6632.2009.04152.x

Tallis, H., Kareiva, P., Marvier, M., & Chang, A. (2008). An ecosystem services framework to support both practical conservation and economic development. *Proceedings of the National Academy of Sciences, 105*(28), 9457–9464.

van Eijk, P., & Kumar, R. (2009). *Bio-rights in theory and practice: A financing mechanism for linking poverty alleviation and environmental conservation*. Wetlands International.

Wainger, L. A., King, D. M., Mack, R. N., Price, E. W., & Maslin, T. (2010). Can the concept of ecosystem services be practically applied to improve natural resource management decisions? *Ecological Economics, 69*(5), 978–987. https://doi.org/10.1016/j.ecolecon.2009.12.011

White, C., Costello, C., Kendall, B. E., & Brown, C. J. (2012). The value of coordinated management of interacting ecosystem services. *Ecology Letters, 15*(6), 509–519. https://doi.org/10.1111/j.1461-0248.2012.01773.x

Wünscher, T., Engel, S., & Wunder, S. (2011). Practical alternatives to estimate opportunity costs of forest conservation. In *EAAE 2011 congress change and uncertainty – challenges for agriculture, food.*

Chapter 6
Biorights in Local Context

Dipayan Dey

6.1 Can Biorights Be Claimed by Locals?

Rights are said to be the ethical principles or legal forms of freedoms or entitlements and are considered as the fundamental rules of the society that determines the allowances or ownerships of people as an individual or group by the existing legal system, societal convention or ethical theory. 'Rights' structure the form of governments, the content of laws, and the shape of morality as it is currently perceived (Wenar, 2021). We would definitely like to see biorights of commons in this light of a normative perception, though we would also explore its viability as an entitlement or claim in the existing socio-juridical framework.

We all know that rights are very fundamental to civilizations as they often make the established pillars of society and culture of any ethnic group and the history of recognition of a right is often associated with the history of societal conflicts. But the global perception of right is unanimously twined with ownership. This can also be otherwise stated that ownership on something can be claimed whenever there is a right over it. In land rights or property rights, possession of that land or property is a direct proof of ownership and therefore becomes a major point in legalizing the claim. In case of rights those are intangible or immaterial in nature, either one must own the proof of its impacts, advantages and brunt or must act to claim that right. For example, any claim for fiscal compensation as maybe in case of disaster relief, damage and loss in natural hazard, man-animal conflict in forest etc. the claimant needs to substantiate the claim with the possession of an evidence of its impact. As in case of flood or earthquake, the compensation claims may need to be substantiated with a proof of residence or possession of ant property in the area impacted. Likewise, claimants of an accident victim must produce medical documents for his insurance claims etc. However, in cases like fundamental rights of citizens it is

D. Dey (✉)
South Asian Forum for Environment, Kolkata, West Bengal, India

D. Dey, J. Roy (eds.), *Biorights*, Studies in Ecological Economics 7,
https://doi.org/10.1007/978-3-030-91503-2_6

expected that the citizen must act responsibly or follow the citizen's charter of responsibilities and that proves to be their ownership of citizen's rights. Therefore, in semantically defining rights we find three subjects, namely an individual or a collection of them, an action or a series of actions and a substance or a property, who are different but not compatible and as well connected by a legal or moral binding or commitment that grants some benefits only to the first subject – an individual or a collection of them. This relationship or bondage of commitment that grants benefit is often termed as entitlement or right. Since rights are benefits or edges, they are commonly connoted as entitlements, wherein any ownership or bond represents a property or material of value to its holder or bearer. Human rights are usually seen as entitlements of persons or individuals to essential goods, of which it can be inferred claims against other persons or against governments and officials (Azevedo & Antonio, 2010).

There is an influent idea about rights that it is a natural or legal bonding between an individual and a property from which specific physical or juridical demands can be derived against a person or a party, whereas a persuasive idea about right suggests that it is a relationship that defines a demand or claim of a person over another, either about an action or any state-of-affairs. Technically, the influent view is called as the entitlement and it considers rights as relationship between something like property or material with someone, who possess it, while the persuasive view is known as claim and it considers rights as a demand of one person over another, of which the other person may be determined or undetermined. For example, the claim maybe of one person against another determined person whose identity is known, as in case of fishing rights between the pond owner and the fisher, else it can be the claim of one person against another undetermined person who represents a group or an organization, as in the case of the claim of a victim attacked or killed by wild animal against the department of forest, being represented by an officer. Thus, a claim or entitlement against any right depends on the concerned parties; likewise it can also be associated with the essentialism or non-essentialism of rights. If driving licenses are beneficial to people, the ability to drive is also a benefit. Saying that someone has a driving license would naturally imply that he can also drive, Rights too are similar in views. Saying that someone has the right also implies that the person is enjoying that right as well. But there would be certain rights that would be beneficial in an essential sense, for example right to water. Without water the existence of the person would be at stake. Such rights would be vital and fundamental and denial of recompense against such rights becomes a juridical offence.

In this pretext, if we look up at biorights, we find a quasi-perfect feature to fit in with the definitions discussed so far. The basic reason is that biorights is neither a claim nor entitlement, rather it is an incentive to motivate voluntary pro-conservation activities towards augmenting adaptive practices and community resilience. On the contrary, biorights strengthen the natural community rights of commons on the habitat as 'right to inhabitancy', right to livelihood, right to access to natural resources etc.

6.2 Socio-political Attributes of Biorights of Commons

Setting up conservation objectives of any local ecosystem largely depends on the participation of the local community and as well its other stakeholders. Whenever, we talk about climate change issues in the local context, we must recognise that the importance of the local agenda, local policies and societal norms are very important in designing a location specific solution. In this perspective if we review biorights, we find that it is not a stand-alone conservation model rather it is an integrated approach and provides a local package for wider conservation and development portfolio. It has to be locally designed place-based intervention that addresses local agenda of the commoners. Let us briefly discuss a case study to understand the context. Among the highlanders of north eastern Indian states, women farmers are often engaged in crop cycle planning and farming of traditional crops like local aromatic rice or buckwheat etc. Now, with the impacts of changing micro-climatic patterns, the production has fallen and farming terraces have been converted to exotic cash crop farmyards like Kiwi or Chinese tomatoes. Though, these crops earn profit but it also causes adverse impacts on the local agro-biodiversity, threats food security of local people, brings in inequity of access to economic and natural resources, affects the social status and livelihood of women farmers as they become out of job and overall sustainability is locally dilapidated. To address this scenario, it is important to engage local communities with a holistic approach. While new financial mechanisms need to be drawn up to sustain local farming practices through market linkages and value chains so that biodiversity could be protected, local institutions too have to be strengthened for evading the impact of climate change on gender equity.

This raises the scope for biorights to provide a package for wider conservation and development in the local context. The relevance of biorights can be well assessed with its components, where a place best intervention addressing the local agenda of the communities attempts to provide a solution in the climate milieu. One such important component is innovative financing system and another is building capacities of the local communities to leverage that system. Therefore, if we consolidate the concept, we find there are significant local drivers and determinants as well that helps to develop the place-based biorights package. Some of these are discussed here in the following passages.

6.3 Local Policies and Law Enforcements

Local policies are often made in the home-grown socio-political context. The policies are mostly designed by political entities who usually develop it on political motivations and vested interests. Therefore, local policies may or may not enable biorights. However, in many cases biorights can trans-create some local policies to community practices that can sustain conservation and development as well.

One good example is in the state of West Bengal in India the municipal affairs department had implemented a policy for the livelihood generation of urban poor. This policy could well be utilised in extending support to the Peri urban wetlanders who survive on the ecosystem services of the peri urban wetland by practicing aqua farming. Later, when these marginal communities were engaged in biorights intervention the same policy helped in developing the innovative finance mechanism. Thus, biorights help in operationalizing policies. This facilitates a great deal of policy hiking from local level to state level, then national level and even regional level interventions. However, when we talk about law enforcement from a policy decision point, we find that the enforcement could be either a top-down process, wherein the beneficiaries are legally compelled to conserve the habitat, or on the other hand it can very well be a bottom-up approach, where community is engaged for enforcing the law. Both these approaches are not mutually exclusive. When we find a law being enforced it helps to identify the illegal usage of resources and the law bans it. The marginal community members, who were using the resources definitely loose some opportunities, which they cannot accrue otherwise. As a part to this end, if this opportunity cost be compensated by incentivizing with some other opportunities or even with fiscal assistance then the local communities can better be engaged directly in law enforcement and as well the strict enforcement of contractual obligations would enable them to stop outsiders, who wants to encroach illegally for resource gathering. Thus, biorights can be a Win-Win situation for conservation and development in any legal framework or action.

6.4 Ownership Rights

As we have discussed earlier the ownership rights of the community members or tenure rights that locally establish secured land rights, can always strengthen any biorights intervention to accomplish better success. The simple reason being that where the land tenure is secured there will be lesser interferences and conflicts, mitigating the risk factors to a large extent. In such cases, long-term management strategies can be planned with the community members much easily then in cases where the land holding has a distributed ownership or it is a vested land with user rights being devolved with government line departments. In case of Ramsar wetlands in eastern and north-eastern India, we have seen that there are multiple stakeholders and line departments who hold the land rights and the beneficiaries could not be mainstreamed as a formal cooperative or business enterprise just because they didn't enjoy the tenure rights of the land. In case of DeeporBeel Ramsar wetland in Guwahati city of Assam, the wetland area was conserved and managed by the local indigenous people but the land belonged to the Department of Forest, whereas in East Kolkata Ramsar site on Kolkata city of West Bengal, the land users were the department of fisheries, department of irrigation, municipal affairs department and even rural local administrations called '*Panchayats*'. In both the cases, local organization of the communities engaged in livelihood activities centring the wetlands, could not be constituted owing to lack of secured land tenures and overlapping jurisprudence of the line departments.

6.5 Aligning Biorights to Local Conservation Approaches

It is important to align any local biorights intervention with other conservation modules already existing with a community-based approach in the locale, where innovative financial mechanisms are included in the intervention design. This helps in tapping the global market economics for ecological purposes. Currently, in the milieu of climate change, several such conservation practices are being developed and piloted in various parts of the world. These are dealt in further details in Chapter No. 4. Some of these have been successful in resource conservation and resource mobilization for conservation as well. For example, the carbon financing mechanism of the Green Climate Fund or Adaptation Fund for least developed countries and small island countries or REDD plus for forest communities and Blue-carbon interventions for conserving the coastal habitats and its inhabitants, are showing promising results in the locale.

In each of these approaches the beneficiaries need to be properly trained to understand the mechanism and such type of capacity building is integral to the biorights paradigm. Biorights, as a neo-economic conservation paradigm, is therefore perceived as an instrument for potential mobilization of economic benefits from the global green markets to the marginal communities at the bottom of the pyramid, engaged in conservation activities. To improvise this, awareness drive among the beneficiaries and stakeholders is essential. In most of the situations there are several scientific assumptions and assessments intertwined with the biorights principle. Unless the beneficiaries and policy planners are not aware of these fundamentals and do not recognize the decision approval system based on these assumptions, the guarantee of success and reliance on the intervention will remain too low to be accepted. Therefore, technology cooperation and handholding during technology transfer becomes crucial for a successful and sustainable intervention. One such example is illustrated here as a case study to understand the constraints of integrating plurality of criteria assessments in biorights to establish it as a mainstream conservation activity. This case study is based on 'Community Water Banking' practiced in the drought prone areas of Eastern India.

In many cases, the payments for compensating the opportunity cost may work as a trade-off between conservation and development. The beneficiaries may assume that conservation efforts are the conduits to get biorights payments, whether or not the conservation objectives are fulfilled through those efforts. Cumulative savings from such payments can further drive them for developmental needs to facilitate alternative livelihood ventures thereby creating a conflict of interest between the stakeholders and the beneficiaries. Ecotourism was established in East Kolkata Wetlands as a pioneering intervention on biorights and the revenues earned compensated the opportunity costs by supporting the fisher's communities in the lean

Community Water Banking[1]

Ecologically water is critical in Purulia, a drought prone district of West Bengal in eastern India, owing to its sloppy terrain and high run-off loss. Though, water demand has grown with scarce supplies in drought-hit areas and its aggressive want for hybrid field crops, but not with such variability as supplies from above or below ground stocks. This has been dealt traditionally so far in this area, as local institutes are not in place to let the stakeholders manage water as a commodity or a public good being shared within an irrigation system. The intervention aims at achieving security of irrigable water in Purulia through right-based entrepreneurial water conservation paradigm for leasing water resources between willing water-rights holders and willing-to-pay users to assure reliable supply of quality water in budgeted volume. This includes water banks built on harvested rainwater for irrigation using solar pumps that extends ecological services like ground water recharge; sustain vine horticulture atop and fisheries as well.

Milestones of the intervention were:

1. *Installation of community water banks at strategic locations for its sustainable management by trained youth*
2. *Participatory crop cycle planning and water budgeting for each water bank.*
3. *Activating user accounts in every bank for Sustainable operation and water service delivery*
4. *Initiating value added alternative farming practices for compensating opportunity costs.*

In cognizance with 'Biorights' of commons that supports incentivizing entrepreneurial conservation through payments from ecosystem services motivated the intervention and accomplished twofold increase in crop yield by water banking along with considerable reduction of water footprints through wise usage and integrated farming practices that assured the second crop. Adaptive management over time and space could be ensured with continuous monitoring and evaluation of progress, while strategic impact assessment appraised major outcomes of the project. It is observed that strengthening local institutions as formal water user's associations towards financial inclusion is essential for accomplishing the intended outcomes. Risk mitigation through risk coverage and better risk spreading will be crucial. However, getting the ReSource award will add unique momentum in the movement and ensure co-funding for dimensional scaling up to attain a sustainable size and resilient system.

(continued)

[1] https://www.equatorinitiative.org/2017/06/05/community-water-banks-for-drought-risk-resilience-in-purulia/

Social and Ecological Challenge

Exponential growth in demand for food has pushed farmers to move from subsistence to commercial agriculture. The water demand has grown gradually, but not with such variability as supplies that had been dealt so far through hydrological scenario analysis. In the climate milieu this doesn't work. Consequently, the de-facto water distribution practice has to change by delivering the proper quantity and quality of water for crops as and when required (Hindustan Times 2016*). Unfortunately, our place of intervention Purulia does not have required legal and organizational infrastructure in place to let the stakeholders manage water as a commodity being a shared or a common asset at different levels of an irrigation system. This societal challenge has amalgamated with ecological water crisis owing to the terrain patterns in Purulia, wherein the run-off loss is very high. Previously, irrigation management practices hardly had decisions about the time, the quantity and quality of irrigation water to apply at each stage of crop growth. Presently in most areas, with or without a crop planted, the irrigation management is conceived to be a system of weekly river water disposal, equitable to the land owned but unreliable and inequitable in terms of quantity and quality. Obviously, the system is too rigid to adjust to the escalating demand for irrigation water. Therefore, as against canal water, the pumped groundwater comes close to meeting the current purpose of irrigation. Within the available know how of farmers, tube-wells too are operated at a guessed duration to meet crop water requirements. If a rigid water delivery system served the purpose to keep famines at bay by having just subsistence agriculture during the nineteenth century, a flexible system is the demand of the twenty-first century to re-engineer an irrigation supply system that is more reliable, adequate and equitably accessible and as well create a market for water stocks below or above the surface. In Purulia, where 80% available water is used for irrigation, it's important to understand the challenge of shifting historic practices to systemic irrigation management. Remoteness has ostracized farmers from modern agro-farming and technology cooperation. Perceptions of collective farming, crop cycle planning on budgeted water or lending water rights, therefore, stumble against several primal societal inhibitions. Though climate aberration in Purulia is multiplying water risks, farmers are still willing to pay more for expensive and inferior quality of groundwater because of the flexibility and control over extraction. Unfortunately this is exploited for political advantages by offering free power or pump. If, one important factor in dejecting water banking is the rigidity, unreliability and inadequacy of surface water at field level, the other is the lack of robust local institutions, which must assure cooperative strength and mutual trust for preparedness against water arrears.*

(continued)

There have been consorted efforts to balance the supply side with the water demand, using traditional methods of check dams on the run off, rain water harvesting and shallow tube-wells. It is estimated that 15 minutes of relatively inferior quality pumped water now costs more than what the farmers pay per acre using better quality canal water for the entire irrigation season. But just because of the flexibility in time and control over the ground water, farmers were willing to pay more for the ground water. This led to an ecological imbalance and many bore-wells were running dry. There have been usable surface storages but the sole proprietorship prohibited its collective usage from a fear of exhaustion. A summary of land use pattern shows that 12–15% of surface area in a village block has water storage facility and 17–23% is arable lands, rest 65% land remains uncultivated due to water shortage. Unfortunately the thoughtfulness of budgeted water use was an absconding idea. Chasing water through run-off reduction pond accumulation and sucking it from the ground remained to be common ways of meeting water demand. Naturally the agrarian economy became as unpredictable in primary production as the Indian rains. The government schemes had labels of vested interests, which the commons understood but could not undermine. Later, drought relief and government subsidies became more desired substitutes for a good harvest. High dams on run-of-the river are not uncommon in Purulia but there have been several points to letdown. Defunct lock-gates, unmapped watersheds and unaccounted ecological impacts owing to reduction in environmental flow are few to mention. Interestingly, these projects targeted tourism prospects around water barrages rather than facilitating irrigation. Community endeavours at family level have ensured reclamation of ponds but it failed to come up as collective activity for obvious reason.

The scheme envisages that water for lift irrigation is a commodity which is truly 'liquid' in form and nature, being sold against an estimate of opportunity cost to partners, who are farmers. The services extend above just irrigating drought prone highlands as water-banks recharge ground water, sustains horticultural vines above the water surface for vegetable production and can be used for fisheries. Depending on the spatial location of lands of the members in this water partnership, the lowlanders can be bankers and highlanders can be investors of the run-off or borrowers in times of need. While the bankers earn incentives accrued from alternative farming and service charges, the highlanders pay user fees and interests for borrowing water rights of others and enjoy equitable access to water resources with guaranteed volume and quality. The interesting part is that partners actually lend or borrow the water rights and use the water services and compensate the opportunity costs thereto. The conservation paradigm is right-based in entirety and promotes sustainable intensification of ecosystem services. The target beneficiaries are all marginal farmers and fishers residing within the watershed territory of the water bank.

(continued)

This territory is decided on participatory water budgeting and cropping patterns to suffice and sustain the usage. The community is handhold for such decisions with modern geospatial support systems. Any partner can play the role of water-banker or water-investor and the decisions on partnership modalities too are participatory. The same partner can also be a borrower of water-rights and use the services, however this borrowed right is non-transferable and must be used within the season. If the partner fails to use or violets the mandates of water bank, he pays a penalty which is more than the opportunity cost. The cash transaction is minimized through virtual banking of water-accounts calculated on opportunity costs as investments required for equitable irrigation through diesel pumps and for this calculation capacity is built amongst the partners. One unique success factor is the inclusive growth and easements of marginal commons, who borrow the water rights without ignominy, and are not exploited over the mounting interests. Endowment based gratis use of water-rights saves economically challenged partners or members in distress as an insurance for risk coverage. The other success factor is that it has led to collective farming, preferably organic agricultural, which is a low-water and no-chemical practice for climate adaptability. This is a better form of agri-business that assures critical production and quality yield to fetch sustainable market prices even for the smallest farmer. Lastly it ensures gender equitable resource allocation which is very significant in indigenous matriarchal communities, as in this case. Active women farmers, who can't migrate as agricultural labours in drought periods can now sustain family farming on their own. Net shown area (NSA) in Purulia is 348,500 ha or 56% of total area. 67% of NSA is upland experiencing high runoff and water scarcity and 73% belongs to marginal farmers having fragmented smallholding. Only 17% is under multi-cropping and rest is mono-cropped for rain-fed cultivation. Present intervention in Chaturmahal, an agrarian tribal village of 305 hectare with 215 households having 950 heads wherein 98.24% are below poverty level and 49.7% are female, achieved water security for 700 farmers by establishing water banks with geospatial mapping for drainage and elevation using ASTER DEM and resource sharing through delivery of measured volume of quality water for irrigation post water accounting and runoff-estimates. Twofold increase in crop yield, 57–65% reduction of water footprints through participatory crop cycle planning, and 28–30% reduction in evaporation loss through integrated vine farming practices has been able to assure the second crop here in last 30 months.

In the proposed area, 98% farming practices are rain-fed agriculture though nearly 65% arable lands stay empty owing to water scarcity. A scoping study gives proximity of 37% exponential growth in merely three consecutive monsoons, if there be water banking. Recent government schemes have initiated fiscal supports to marginal farmers in the area but quality and

(continued)

volume of water resources cannot be assured thus and therefore failed to augment yield. Collective action and participatory planning for resource allocation and budgeted use in the proposed water-banking model have augmented resilience wherein 82% of the inhabitants have shown concurrence within the small intervention area. Integrated watershed management and conservation perspectives based on basin ecology in defining watersheds are the strengths of the intervention that guarantee a scaling up not only in all drought districts of the state but in the whole agro-ecoregion spanning six more states of eastern India. Government has launched copious subsidized schemes for solar irrigation to promote clean energy in farming. Water availability can only enable such facilities assuring potential scaling up of water banks. Foreign Direct Investments in rural water and sanitation through corporate responsibilities have shown multi fold increase in India, whereas 'National Water Mission' is encouraging entrepreneurial solutions in rural water management. These are conducive for quicker scaling up too.

Water bank is an establishment for the custody of irrigation water, which will change from private enterprise to user managed entity. Shifting demand and supply patterns at user end will change water delivery from open used arrangement to consensus-based irrigation management system. In climate milieu, increasing frequency and intensity of rainfall, floods and droughts, will make water banking system a handy tool for a reliable supply side of the water management, pronouncing steady scale up. Minimal landholding is 1 Ha for water bank beneficiaries and average size of water bank needs to be minimum 5% of watershed area with 3.5 m depths. Micro-irrigation & Zero-tillage saves up to 35% water for second cropping. Dry weather pond capacity preferably is above 85,000 Cu Mt to sustain fisheries and lift irrigation. 60% of water from water-bank can be withdrawn at the maximum. 3Kw solar unit with 2HP pump and 5 drip-irrigation set is enough for 20 hectares with average 2–7 feet elevation.

periods. As the returns on investments from ecotourism started gaining volume, the community decided to create infrastructures for facilitating tourists that ultimately caused alteration of characteristics of the wetlands, which is against the Ramsar norms in entirety and finally had to be stopped. In such cases, built capacities and awareness campaigns can bring back the synergy once again. For this, it is important to organise the communities into special groups. Usually in a marginal rural community there are diverse groups wearing in ethnicity e socioeconomic status cultural practices education etc. In the social dynamics there always exists several conflicts between such diverse groups and it can utterly disturb the bi rights intervention making it difficult to accomplish conservation objectives. To address this issue, specialised interest groups based on nonspecific abilities or societal features can be created leaving no-one behind. These are termed as joint liability groups, who would basically manage and maintain the contrivances in biorights intervention. The advantage of such groups is that the diversity is thawed and conflicts are inevitably resolved facilitating inclusive growth and community level sustainability.

6.6 Biorights and Community-Based Natural Resource Management (CB-NRM)

Community-based Natural Resource Management (CB-NRM), has a special advantage of transforming policy from paper to practice. If any developmental activity that involves natural resource management, is designed in participation with the community then the policies derived becomes more realistic in the local context and therefore acceptable to the community. It becomes easy to convince the community members to adopt such practices, whatever may be the opportunity cost, with minimal chances of conflict or confrontations. However, this did not happen so far as because the opportunity costs do not get compensated easily or locally and political interests drive the show keeping the commoner's side in absolute dark. Since the ideas are not paying off, the interest of the communities engaged in natural resource management die out putting the natural resources at bay. When we talk about biorights this bottleneck is addressed more equitably and settled at the community ecosystem interface.

6.6.1 Community Groups in CB-NRM

Usually, to facilitate microfinance different types of community groups like self-help groups, women empowerment groups etc. are being formed and the groups are subsequently linked with local banks by opening a group account. Post that the group members collectively invest a nominal amount in the account and the savings are used for internal lending on a token interest. The profit slowly accumulates and the group savings account increase. Seldom the group gets facilitated with credit linkage schemes or any other income generation facility. Quite often it is found that conflicts arising from lack of financial literacy and awareness, transparency in transacting and inequity render the groups to be defunct. In case of biorights, the compensation or the incentive is deposited in these group level bank account, capacities in financial literacy is built among the members and the group dynamics is mentored, so that the irregularities and gaps in documentations can be evaded. It is observed that these activities help in regularizing banking activities and savings volume grow good enough to reinvest in natural resource intensification like plantation, restoration and reclamation. Investment in de-silting wetlands and embankment stabilization with energy cropping are few such examples of good investments. This revives the primary productivity from the habitat, restores the ecosystem services and thereby earns better profit for the marginal fishers and farmers to encourage entrepreneurship in the community. The same capital can also be used as a seed fund for alternative livelihood opportunities like food processing, packaging, etc. and capacities are built thereto. This can be extended especially for community women those who are compelled to work in informal sectors like waste workers or household help in urban settings. In some of the successful biorights intervention a part of the capital is also seen to be invested in paying premium fees for group micro-insurance schemes for the members in extending risk coverage for the group. Therefore, the biorights groups are more interested in in community based natural resource management compared to other societal groups and stay functional through collective efforts to enable community saving schemes. It is interesting to note that few such small groups have developed into community cooperative and micro-finance groups thereby strengthening the local institutions at the community level and augmenting community resilience in the climate milieu.

6.6.2 Conservation and Development in Local Political Agenda

Though it is well acknowledged that conservation and development are inextricably coupled, yet conservation objectives have yet not been included in the local political agenda in its true perspective, especially in the developing countries. The dichotomy between conservation and development priorities originates from the upper tier of political leaderships, who are not connected organically to the local problems. Development objectives are often superficially articulated in the political agenda to keep the community in illusion of some material benefits and facilities and ad hoc commitments for politically vested interests. In this gamut of commitments, the conservation issues appear either as a bottleneck problem like scarcity of water in the locale or defunct sewerage management system etc. else it comes as a criticality or threat like man-animal conflict and environmental hazard mitigation. To be politically correct the leadership addresses it sketchily and often the suggested solutions are temporary and non-sustainable. This socio-political disparity between so called development, which often relates to creating infrastructures for urbanized facilities and conservation issues comes from a political apathy towards environmental conservation that percolates from the top leaderships to the local level. The reason behind this disparity seems to be an apprehension of opportunity cost and as well knowledge gap owing to ignorance about environmental processes and functionalities. On the other hand, the green sector too, has failed in marketing its bouquet of benefits making it politically lucrative. However, there are instances, where the conservation issue had been politicized and mass socio-political movements have been drafted out of it. The '*Narmada bachao*' movement is one such example, where socio-political movement raised by the local people on the conservation issues of riverine ecology of river Narmada owing to construction of dams that was supported by the World Bank and Government of India, was put on hold (**Plate on NBA**). Conservation issues gather such momentum usually when it relates to life and livelihood security of people in the immediate context.

In this backdrop, when we see biorights, we find there are two interesting facets in it that is compensation of opportunity costs of commons and building capacities for technology cooperation that prepares the communities and the stakeholders to bridge this gap between conservation and development and bring it at the centre of political agenda. Here is a small schematic representation to illustrate how biorights can be placed for its inclusion in the local political agenda.

References

de Azevedo, M. A. O. (2010). Rights as entitlements and rights as claims. *Veritas (Porto Alegre), 55*(1). https://doi.org/10.15448/1984-6746.2010.1.7327

Hindustan Times. (2016). https://www.hindustantimes.com/india-news/collective-farming-water-budgeting-purulia-s-weapons-against-drought/story-jbttY76cUHA3QLIoNSrJzH.html

Wenar, L. (2021). Rights. In E. N. Zalta (Ed.), *The Stanford Encyclopedia of philosophy*. https://plato.stanford.edu/archives/spr2021/entries/rights/

Chapter 7
Biorights in the Climate Milieu

Dipayan Dey, Sourabh Kumar Dubey, Malancha Dey Roy, and Udo Censkowsky

7.1 Introduction: Climate Change and Environmental Justice

Climate change is posing major threat to 'Right to Life' that has been globally acclaimed as the most important and fundamental human right. The fact that nearly a million premature deaths are being linked globally to climate crisis is evidence how climate change is looming over the realization of this fundamental right. It is a matter of fact that with the rising temperature and climate complicacies that includes extreme weather events like El Nino Southern Oscillation (ENSO) or Indian Ocean Dipole (IOD) etc. generating heat waves, floods, droughts, wildfires, water-borne and vector-borne diseases, malnutrition and air pollution the challenges on human rights will be escalating beyond limits (Thirumalai et al., 2017). IPCC reports (AR6) have already indicated that the pressure on water resources, which makes the most essential ingredient for life on the planet, will be maximum. This would entail reduction in per capita availability of safe water defying the right to water and sanitation and challenging SDG No.6, whereas water usages patterns will compel urbanites to ration consumptions as well farmers to adopt low-water intensive cropping practices thereby widening the urban-rural divide. At 2 °C, 100 million more people are forecasted to face water insecurity (Boretti & Rosa, 2019).

Climate impact is the biggest global threat in the twenty-first century that could reverse five decades of progress and preparedness in global public health as it endangers the right to food and nutrition and ambient air for health wellbeing

D. Dey (✉) · M. D. Roy
South Asian Forum for Environment, Kolkata, West Bengal, India

S. K. Dubey · U. Censkowsky
Blue Census, Weilheim, Germany
e-mail: u.censkowasky@blue-census.com

D. Dey, J. Roy (eds.), *Biorights*, Studies in Ecological Economics 7,
https://doi.org/10.1007/978-3-030-91503-2_7

(https://www.thelancet.com/journals/lancet/article/PIIS0140-6736(15)60854-6/fulltext). The recent pandemic of COVID-19 has proved it all. The pandemic and as well other zoonotic diseases have been cross-linked to climate emergency furthering the challenge against right to health. This may lead to increased malnutrition, stunting, wasting, allergies, injuries, incidences of respiratory and cardiovascular disease, and societal psychosis as well. Dengue and Nipa Virus, for example, has been one among the most rapidly spreading vector-borne disease that has showed nearly 30-fold increase in global disease incidences and this spread has been largely attributable to climate change.

Food and Agriculture Organization of the United Nations in its 2018 report has mentioned that climate variability and extremes are among the key drivers behind the recent uptick in global hunger and one of the leading causes of severe food crises. The cumulative effect of changes in climate is undermining all dimensions of food security, including food availability, access, utilization and stability (http://www.fao.org/3/i9553en/i9553en.pdf). United Nations Human Rights special procedure report A/74/161suggested that by 2050 more than hundred and fifty million people will be displaced as environmental refugees owing to rising sea levels, coastal erosion, storm surges, salinization and other climate impacts and shall have to be relocated keeping the human rights concerns in the objectives of relocation. Several states including Fiji, Kiribati, Maldives and Tuvalu in entirety are at risk of becoming uninhabitable and may have to face this critical problem in the near future. (https://wedocs.unep.org/bitstream/handle/20.500.11822/30158/Safe_Climate_Report.pdf?sequence=1&isAllowed=y) This raises a new domain of concern that is right to climate justice. Though 155 member states of UN Organization has recognized healthy and safe environment as a matter of right to the inhabitants but at the same time the efforts to address it in right's perspective is very inadequate.

7.1.1 Equalizing Climate Change Policies Using Rights to Be Informed

It is a global undertaking that the governments have to enhance the adaptive capacity of vulnerable populations in mitigating climate change and it is obligatory for them to take effective measures to prevent foreseeable loss of life. This includes but are not limited to establishing, implementing and enforcing laws, policies and programmes to fulfil their citizens' rights and as well preventing the potential violation of those rights by third parties, especially commercial organizations, as well.

The UN Human Rights Office recommends that commercials and corporations must have their concurrence and compliances to the 'Guiding Principles on Business and Human Rights' in adopting human rights policies, conducting human rights due diligence and remedying human rights violations in working environments and work to influence other actors to respect human rights in the climate milieu. Such right-based approaches are expected to motivate ambitious actions to facilitate the marginal communities highly vulnerable to climate impacts as well makes it obligatory for the governments and corporations to encourage participation and partnership in designing climate adaptive solution and empower the local inhabitants.

Reviewing and exploring the facets of human rights in the milieu of climate change by several human rights organizations, governments and states, as well UN Human Rights organs and international courts of justice have revealed that climate change impacts are threatening a broad range of fundamental human rights and the states, governments, private actors and all other related organizations have extensive human rights obligations and responsibilities thereof.

United Nations Human Rights Council in its Resolution 26/27 (https://documents-dds-ny.un.org/doc/UNDOC/GEN/G14/083/51/PDF/G1408351.pdf?OpenElement), emphasized that climate change is an urgent global problem requiring a global solution. It requested for global alliances and other UN organs like UNFCCC in order to support national efforts for the realization of human rights affected by climate change related impact. The Council affirmed that "human rights obligations, standards and principles have the potential to inform and strengthen international, regional and national policymaking in the area of climate change, promoting policy coherence, legitimacy and sustainable outcomes." The councils panel held on 6th March 2015 to discuss human rights and the challenges posed by climate change, especially to the vulnerable marginal communities had submitted extensively deliberated recommendations complemented by human rights commentary and analysis from all cross sections of the society, organizations, CSOs and academia to the 21st Conference of Parties in Paris. The discourse in COP 21, Paris 2015, stressed elaborately on the legal basis of violation of specific human rights being affected by climate change as well the factual basis for human rights violations allegedly caused by climate change and duly recommended for a rights-based approach to address these violations owing to climate change impacts.

7.1.2 Significance of Human-Rights Approach in Environmental Justice

Human rights assure universally acclaimed legal guarantees towards protecting against actions and denials that restrict fundamental freedoms and entitlements of human beings. The law thereto directs all stakeholders including governments and corporations to respect, promote, protect and fulfil fundamental rights of human beings those are universal and are based on inherent dignity and equity amongst all human beings. These laws therefore have to be equal, indivisible, interrelated and interdependent, and cannot be waived or taken away under any circumstances. Additionally, the legal protection imposes obligations on States and State actors in relation to actions and denials of these rights. Climate change, being anthropogenic in origin, endorses our accountability to evade the scopes and chances of its profound impacts on a wide variety of human rights. These essentially includes, rights to life, self-determination, development, food, health, water sanitation and shelter. Therefore, the human rights framework also guides the global efforts to ensure the rights to participation and information, transparency, accountability, equity, and non-discrimination in community-based mitigation and adaptation to combat climate change. The Fifth and Sixth Assessment Report (AR5 & AR6) by the Intergovernmental Panel on Climate Change (IPCC) has pronounced without any iota of doubt that the anthropogenic causes of climate change is well established and people who are socially, economically, culturally, politically, institutionally or otherwise marginalized are especially vulnerable to climate change and also to some adaptation and mitigation responses.

While addressing the Human rights panel, former UN High Commissioner for Human Rights, Mary Robinson, observed that climate change is probably the greatest human rights challenge of the twenty-first century and other panellists described climate change as an ethical and moral challenge of a global dimension that needs a human rights approach to solve. Recognizing the significance of human rights in the climate change context, it was agreed in the conference of parties to the UNFCCC to fully respect human rights in all climate change-related actions. However, this was also resolved in the COP forum that despite there were clear human rights implications, the efforts of the international community were substantially inadequate in preventing climate change impacts on rights and entitlements of people, especially the marginal communities. Furthermore, many of the mitigation and adaptation efforts deployed till date have had counterproductive human rights impacts, particularly on the them. The wholesome efforts and interventions have fallen short of ambition that was essential in abating the negative human rights impacts of climate change in fulfilment of the obligations of all stakeholders and duty-bearers. Therefore, an integrative approach of clubbing human rights in climate actions needs to be enabled to improve mitigation and adaptation strategies. This will not only be more effective and inclusive, it will also address right-based climate policy, including participation, access to information and access to environmental justice.

Any rights-based approach in persuading environmental justice has to address cross cutting issues in societal, cultural, political and economic domains, while empowering communities in vulnerable situations. The ethnobiological paraphernalia and traditional knowledge is equally important for including the indigenous communities in the purview of environmental justice. This substantially contributes to climate change policies, making them less myopic and more approachable, being sensitive and collaborative in the interest of the community beneficiaries of the law. The apex bodies on human rights and its definers like UN Charter, the Universal Declaration of Human Rights, the International Covenant on Economic, Social and Cultural Rights (ICESCR), and the UN Declaration on the Right to Development have made it very clear that with a synchronized effort of both individual action and international cooperation the State human rights obligations can't be addressed justifiably. ICESCR has prompted that States should 'take steps, individually and through international assistance and co-operation, especially economic and technical, to the maximum of their available resources, with a view to achieving progressively the full realization of rights recognized in the present Covenant'. The Declaration on the Right to Development further emphasized that "all human beings have a responsibility for development, individually and collectively... and they should therefore promote and protect an appropriate political, social and economic order for development". The UN Guiding Principles on Business and Human Rights confirm that businesses also have human rights responsibilities. They reaffirm that all responsible actors should be held accountable for the negative impacts of their activities and that all actors share responsibility for remedying these impacts. As a part to this end, if we review once again the biorights paradigm, where the right to get compensated for the opportunities lost in protecting the habitat and perusing conservation objectives, we may claim that the basic human rights principles of equality and non-discrimination or the principle of 'Leave No One Behind' in the climate milieu is being addressed in biorights of commons for remedying the disproportionate impacts of climate change on the most marginalized communities to ensure that climate actions benefit persons, groups and peoples in vulnerable situations and reduce inequalities.

The lopsided impacts of climate change on communities in vulnerable situations defies climate justice, impartiality, equity and equitable access to remedy. The Universal Declaration of Human Rights or the International Covenant on Civil and Political Rights (ICCPR) and other human rights instruments strongly advocates for equalizing such disproportionate impacts. The Human Rights Council panel and COP 21 forum had repeatedly called for climate justice and immediate action plans to mitigate and adapt to climate change interventions most equitably at the community-ecosystem interface. The principle of equity, including intergenerational equity, has been specifically recognized in the UNFCCC forum, which calls for all parties to "protect the climate system for the benefit of present and future generations of humankind, on the basis of equity and in accordance with their common but differentiated responsibilities and respective capabilities". Despite that, the 6th IPCC report (AR6) clearly describes that the commitments of the governments and the State to the UNFCCC led global platform have so far not been successful in providing and ensuring adequate mitigation and adaptation measures thereby delimiting adverse effects of climate change on human rights, the economy, public health and the environment. The human rights principles of transparency, participation and accountability have an important role to play in biorights as well. Participation is a basic human right in itself and a precondition or catalyst for the realization and enjoyment of other human rights. In order to ensure sustainable development and appropriate mechanisms for mitigating and adapting to climate change, the ensuing COP26 Glasgow outcome should reflect human rights principles and a rights-based framework to shape both the substantive commitments of parties and the processes by which they are agreed and carried out.

7.2 Integrating Human Rights in Climate Action

Applying rights-based approach to policy and development as adopted in the Declaration on the Right to Development can integrate Human-rights in climate-change related actions as agreed to in the UN Common Understanding of a Human Rights-Based approach to Development Cooperation (https://unsdg.un.org/2030-agenda/universal-values/human-rights-based-approach). Universality and inalienability, interdependence and interrelatedness, indivisibility, non-discrimination and equality, participation and inclusion, accountability and the rule of law are the key human rights principles that the UN Common Understanding emphasizes to outline a conceptual framework for development, keeping to international human rights standards and to promote and protect human rights in all development activities. A rights-based approach, therefore, seeks to redress discriminatory practices and unjust distributions of power, while analysing obligations, inequalities and vulnerabilities of the subjects. It anchors plans, policies and programmes and corresponding obligations established by international law in a system of rights.

The essential attributes of a human rights-based approach in the climate crisis is thus portrayed:

(a) As policies and programmes are formulated, the main objective should be to fulfil human rights.
(b) The rights-holders and their entitlements must be identified as well as the corresponding duty-bearers and their obligations in order to find ways for strengthening the capacities of rights-holders to make their claims and of duty-bearers to meet their obligations.
(c) Principles and standards derived from international human rights law – especially the Universal Declaration of Human Rights and the core universal human rights treaties, should guide all policies and programming in all phases of the process.

A human rights-based approach should be integrated in any climate change adaptation or mitigation measures, such as the promotion of alternative energy sources, forest conservation or tree-planting projects, resettlement schemes and others. Affected individuals and communities must participate, without discrimination, in the design and implementation of these projects. They must have access to due process and to remedy if their rights are violated. During the panel discussions, the importance of a rights-based approach to climate change was raised frequently. Martin Khor, Executive Director of the South Centre, observed that such an approach demands the inclusion of climate justice, equity, respect for human rights as well as international cooperation and solidarity. These factors are essential in supporting developing countries, including through finance and technology. According to Mary Robinson, a human rights-based approach to development should focus on fulfilling for all persons the minimum conditions necessary for a life of human dignity. Faced with climate change, persons in vulnerable situations must have their rights protected, obtain access to measures of adaptation and resilience, and receive the support of the international community. A rights-based response should also maximize inclusion, participation and equality Critically, it is not enough to simply focus on ensuring that climate actions respect human rights. A rights-based approach requires States to take affirmative actions to respect, protect, promote and fulfil all human rights for all persons. Failure to prevent foreseeable human rights harms caused by climate change, or at the very least to mobilize maximum available resources in an effort to do so, constitutes a breach of this obligation. Human rights obligations apply to the goals and commitments of States in the area of climate change and require that climate actions should focus on protecting the rights of all those vulnerable to climate change starting with those most affected. Human rights principles articulated in the Declaration on the Right to Development and other instruments call for such climate action to be both individual and collective and for it to benefit all persons, particularly the most marginalized. The UNFCCC further elaborates upon the need for equitable climate action calling for States to address climate change in accordance with their common but differentiated responsibilities and respective capabilities in order to benefit present and future generations. State commitments therefore require international cooperation, including financial, technological and capacity-building support, to realise low-carbon, climate-resilient, and sustainable development, while also rapidly reducing greenhouse gas emissions. Only by integrating human rights in climate actions and policies, and empowering people to participate in policy formulation, can States promote sustainability and ensure the accountability of all duty-bearers for their actions. This, in turn, will promote consistency, policy coherence and the enjoyment of all human rights.

The Human Rights Council and its Special Procedures Mechanisms have been actively involved, since 2008, in deliberating on the human rights impacts of climate change. The Council has held two panel discussions on human rights and climate change, and have come up with the following resolutions on human rights and climate change till date:

1. Resolution 7/23 (March 2008): The Council expressed concern that climate change "poses an immediate and far-reaching threat to people and communities around the world" and requested OHCHR to prepare a study on the relationship between climate change and human rights (A/HRC/10/61).
2. Resolution 10/4 (March 2009): The Council noted that "climate change-related impacts have a range of implications, both direct and indirect, for the effective enjoyment of human rights …" and that such effects "will be felt most acutely by those segments of the population who are already in a vulnerable situation …"
3. Resolution 18/22 (September 2011): The Council affirmed that human rights obligations, standards, and principles have the potential to inform and strengthen international and national policy-making in the area of climate change, promoting policy coherence, legitimacy, and sustainable outcomes. They called for a seminar to address the adverse impacts of climate change on the full enjoyment of human rights and a summary report of the seminar (A/HRC/20/7).
4. Resolution 26/27 (July 2014): The Council emphasized the need for all States to enhance international dialogue and cooperation to address the adverse impacts

of climate change on the enjoyment of human rights including the right to development. It called for dialogue, capacity building, mobilization of financial resources, technology transfer, and other forms of cooperation to (12 The report of the 2010 Social Forum is contained in A/HRC/16/62.) facilitate climate change adaptation and mitigation, in order to meet the special needs and circumstances of developing countries.

5. Resolution 29/15 (July 2015): The Council emphasized the importance of continuing to address the adverse consequences of climate change for all and called for a panel discussion and analytical study on the impacts of climate change on the enjoyment of the right to health. The Council has also addressed the issue of the impact of climate change in the framework of its work on human rights and the environment, particularly resolutions 16/11, 19/10, 25/21 and 28/11.

Human Rights Council has recognized in these resolutions that the impact of climate change is a looming threat on human rights and it refers to a global problem that needs a global solution urgently. With reference to the context of discussions in UNFCCC and the post-Paris development agenda 2015, the council recognized the importance of addressing human rights at length in its debates, studies and activities to the sessions of the Conference of Parties to the UNFCCC. The Council clearly pronounced that "human rights obligations and commitments have the potential to inform and strengthen international and national policymaking in the area of climate change". The outcome document of the 2012 United Nations Conference on Sustainable Development "The Future We Want" indorsed the significance of human rights for achieving sustainable development. UN High Commissioner for Human Rights had also emphasized on the responsibilities of all States, prior to the conference, to ensure full coherence between efforts to advance the green economy on one hand, and their human rights obligations on the other and that note was openly circulated to all Permanent Missions in New York and in Geneva. OHCHR has produced a number of reports on the effect of climate change on the enjoyment of human rights in addition to this advocacy. Mandate holders of the special procedures' mechanisms of the Human Rights Council too, have regularly addressed the human rights impacts of climate change in the course of implementing their mandates. In 2014, a Special Rapporteur has been particularly engaged on human rights and the environment and issued a Climate Change mapping report that describes statements made by human rights bodies regarding human rights that are threatened by climate change and human rights obligations related to climate change. This report highlights the work of the special procedures' mechanisms, the human rights treaty-bodies and other relevant bodies related to climate change. There have been several joint advocacy efforts too, on human rights and climate change, by special procedures mandate holders. These include:

(a) An Open Letter dated 17 October 2014 to State Parties to the UNFCCC from 28 Special Procedures mandates that urges the UNFCCC "to adopt urgent and ambitious mitigation and adaptation measures to prevent further harm" and to include in the 2015 climate agreement a commitment that "the Parties shall, in all climate change related actions, respect, protect, promote and fulfil human rights for all." It reaffirmed that climate change as one of the greatest challenges of our generation, highlights its disproportionate effects on disadvantaged, marginalized, and excluded person, groups and peoples, and urges the incorporation of human rights obligations into climate change negotiations.

(b) During World Environment Day 2015, a joint statement of parties drew international attention that the devastations even at 2 °C increase in average global temperature would crucially challenge human rights and the Special Rapporteurs urged climate negotiators to reiterate on the obligations that the law places on States to protect and promote human rights.

(c) Another report was also prepared by several Special Rapporteurs for the Climate Vulnerable Forum, from twenty especially vulnerable countries to the effects of climate change on human rights. This report even warned that an increase of just 1.5 °C temperature will threaten human rights, challenging the level of ambition of the current global temperature goal.

It is imperative, therefore, that in order to realise the commitments of the States to safeguard and uphold human rights for all in the face of climate change challenges and its impacts, the States must act immediately for collective actions. Principle 15 of the Rio Declaration on Environment and Development calls for States to adopt a precautionary approach to environmental harms and act "when there are threats of serious or irreversible damage" even in the absence of full scientific certainty. In the case of climate change there is no uncertainty and its damages are unfathomable. Thus circumstanced, the only viable solution is that urgent preventative action is needed and that too has to be immediate, coordinated climate action for effectively mobilizing resources to mitigate and adapt to the change while protecting the rights of people affected by. The need for such action is supported by robust and empirical scientific decisions support systems, including the most recent report of the IPCC, which overwhelmingly confirms that humanity is headed down a dangerous path towards a warmer, more climate volatile, and less secure world. The growing political consensus in favour of urgent climate action, demonstrate that the time for concrete, effective climate action is now. Since climate change directly contributes to the violation of human rights, States have an affirmative obligation to take measures to mitigate climate change, to prevent negative human rights impacts, and to ensure that all persons, particularly those in vulnerable situations, have adequate capacity to adapt to the growing climate crisis. Integrating human rights obligations in climate actions will improve outcomes by providing concrete measures to protect people from the harms of climate change. There is a growing body of evidence that a human rights-based approach will lead to more sustainable and effective results in climate action (http://www.wri.org/sites/default/files/securingrights_executive_summary.pdf).

A human rights perspective offers guidance for addressing climate through a broader lens that encompasses the economic, social, cultural and political dimensions of climate change. It can also reduce arbitrariness in climate change programs and objectives, and ensure that climate action benefits those who are most in need. The integration of human rights in climate action will promote significant involvement from UN bodies and mechanisms mandated to promote and protect human rights, and from civil society actors bringing additional resources to confront the challenge of climate change. In spite of this, concrete efforts at effective climate change mitigation and adaptation measures have been limited. In the course of the panel discussion, President Anote Tong posed the question "how much global effort are we putting into providing concrete, tangible action to help the vulnerable whose security and survival is very much on the line?" The answer so far has been not enough. In order to address this failure, panellists and participants made several recommendations to help effectively integrate human rights considerations in climate adaptation and mitigation efforts and to spur the urgent action that is so critically needed to prevent climate change from becoming a human rights catastrophe.

Recommendations from the full-day panel discussion on climate change in Paris COP 21, 2015 (https://www.ohchr.org/Documents/Issues/ClimateChange/COP21.pdf) is being quoted here below.

(a) Appointment of a Special Rapporteur on human rights and climate change to identify good practices, strengthen accountability mechanisms, engage with the Secretariat of the United Nations Framework Convention on Climate Change and further explore the links between climate change and human rights;
(b) Thorough integration of human rights obligations, standards and principles in international and national climate policy and action, including COP 21;
(c) Participation of all relevant stakeholders in the development and implementation of climate change policies and action;
(d) Mitigation and adaptation efforts that place people at the centre, are gender sensitive, and ensure the rights of persons, groups and peoples in vulnerable situations, including women, children, indigenous peoples and the poor. These could be informed by the use of impact assessments to ensure that climate actions benefit those facing the greatest risks;
(e) Improved regulation of the private sector in order to mitigate their contributions to climate change and ensure their respect for human rights in all their actions;
(f) Enhanced cooperation based on principles of equity and fairness to ensure adequate funding and research into adaptation measures to help the poorest countries and those persons, groups and peoples most at risk;
(g) Intensified research and development into renewable sources of energy and energy conservation to reduce the emissions intensity of growth;
(h) Equitable access to technology, including, if necessary, the lowering of intellectual property standards and facilitation of technology transfer;
(i) The creation of a climate justice fund;
(j) Protecting and utilizing traditional knowledge of local communities and indigenous peoples to support the effective use of resources for agriculture and forestry;
(k) Titling the lands and territories of indigenous peoples and other communities in order to support preservation of forests and improved carbon storage;

(l) Utilizing community-based climate monitoring to reduce monitoring costs and enhance early warning systems;
(m) Improved capacity-building, development assistance, innovative finance and clean development mechanisms;
(n) Improved communication between climate change negotiators and human rights experts including through increased participation in the Geneva Pledge;
(o) Consideration by States of the linkages between human rights and climate change during the universal periodic review;
(p) Increased scrutiny of the impacts of climate change on human rights by Human Rights Council special procedures mandate holders;
(q) A special session of the Human Rights Council on human rights and climate change; and
(r) The creation of a legal instrument to protect the rights of climate displaced people

Recommendations from the panel discussion share a common objective of better integrating human rights in climate action with a view to respect, protect, promote, and fulfil all human rights for all. With negotiations of the outcome document of COP21 currently under way, this objective is critical and States have the opportunity to make considerable progress. From a human rights perspective, the new agreement and other efforts to mitigate climate change and adapt to its impacts should specifically address the human rights impacts of climate change and be consistent with existing human rights agreements, obligations, standards and principles.

7.3 Nature-based Solutions and Biorights in Climate Crisis

It has now been proven that nature-based solutions can help us to protect our life and livelihood from the impacts of climate change by slowing down father warming supporting biodiversity conservation and sustainability intensifying ecosystem services. However, rigorous assessment of the nature-based solutions are required to find the potential that they have in providing us the intended benefits for sustainable development. There are concerns over the reliability and cost-effectiveness of these solutions compared to the engineered alternative usually taken for developmental projects. It is also important to find the degree of resilience that they have to combat the impacts of climate change. If the climate mitigation policy encourages nature waste solutions that has low biodiversity value for example afforestation with non-native monoculture or anthropogenic encroachment in in sensitive ecosystems like wetlands and mangrove forests, then there can be trade-offs arising from such solutions. There are higher chances of maladaptation especially for the rapidly changing world where biodiversity-based resilience and multifunctional landscapes are key elements for sustainability.

In the course of our discussion, we will try to highlight the significance of nature-based solutions in climate policy especially the potential of such solutions in climate change adaptation and mitigation. We will discuss the barriers in implementing such solutions at community ecosystem interface. The various are mainly related to two challenges in the financial and governance of the intervention when long term sustainability is sought. This discussion will also evaluate the potentials of scaling up nature waste solutions and highlight the research and information gaps in the same. In the current social economic scenario decarbonization of the economy and increasing tendency of climate policies towards Greenhouse gas removal demands for neutral evaluation of climate solutions by the natural and social scientist so that empirical evidences can be used for or decision making and policy planning. It is therefore important that's nature-based solution and Ecosystem best adaptations must tackle both the climate and biodiversity crisis as well contribute to sustainable development. This will require systemic changes in implementation design approach and policy planning.

There are three significant challenges of the anthropocene namely mitigation and adaptation to climate change biodiversity conservation and ensuring human wellbeing through sustainable development. Addressing these interdependent challenges would need a multi criteria decision support system along with multi-stakeholder ship failing which nature-based interventions can invite negative consequences and an intended result. The seventeen sustainable development goals propounded by the United Nations and its organs has set the development agenda with a singularity to ensure inclusive and sustained development leaving no one behind in the climate milieu. However, they are interdependencies of these seventeen social, environmental and economic goals that distinctly describes synergies and trade-offs between them. There is little evidence that a balanced action of implementing these goals are happening in practice. The 6th IPCC report (AR6), has clearly explained the failure to stabilize and adapt to climate changes or protect biodiversity. This has been exacerbated by the fact that it is issues are now being handled separately though they are deeply interrelated and have common drivers and determinants. At this point, the potentials of nature-based solutions in concurrence with biorights of commons, make a new possibility for addressing climate crisis.

The new age solutions to face social challenges and economic crisis in the backdrop of climate change are arising as an integrated approach that can counterbalance the trade-offs and synergies between sustainability development goals as because these solutions have relatively low cost of implementation while they can deliver multiple additional benefits for both people and nature. Few simple examples, like restoring peri-urban wetlands can assure the recharging of groundwater for the downstream communities and as well saves the urban inhabitants in the vicinity from flooding. Similarly, restoration of a natural forest that can increase carbon sequestration, also protects forest biodiversity for the life and livelihood of the local and global communities as well. Consequently, the Global Assessment of IPBES and Global Adaptation Commission recommended the inclusion of such sustainable solutions in the UN Climate Action Plan. The World Economic Forum (WEF), while specifically recognising the economic risks posed by biodiversity and habitat loss in its Global Risk report 2019, emphasized the need for nature-positive business solutions. It is evident that the new age solutions are now being viewed as a way to reconcile economic development with ecosystem conservation and as well means to diversify and transform business in enabling sustainable development. At this point, the interventions aligned with biorights provides a unique solution that can be termed as a neo-economic conservation paradigm. In this context we find that nature-based solutions or ecosystem-based adaptations, surface in the policy agenda as a panacea for climate change adaptation and mitigation, whereas we expect them to support the ecosystem services as well. Therefore evidence-based formulation and design approach in formulating these interventions games importance for maximizing the synergies and reducing the trade-offs. It can be stated that in delivering such type of a solution we need to overcome three key barriers, (a) measuring the effectiveness of the solution, (b) mobilizing investments for its implementation and (c) overcoming governance challenges. This explains the need for systemic institutional change that can help in developing a more holistic design with a robust monitoring and evaluation system that can fully incorporate the multiple benefits of such solutions (Seddon et al., 2020).

The latest IPCC report reflects the importance of natural ecosystems, both as sources and sinks for greenhouse gases, as it states that all the scenarios limiting climate change to 1.5 degree Celsius depends directly on the land use change mitigation methods as well decarbonization of the economy. Different land use activities like agriculture, forestry, pastures or built areas together accounts for nearly 23% of total greenhouse gas emissions from human activities. Out of these, the most effective emissions are from deforestation, which gets partly offset by afforestation and reforestation. Though, the 6th IPCC report indicates that there has been a sharp reduction in the sinks of greenhouse gases, it has been also indicated in the report that conservation of terrestrial ecosystem and improvement in agricultural practices have immense potential in decreasing emission sources and increasing sinks for greenhouse gases. This can be further enhanced if such activities are willingly partnered by the stakeholders and community beneficiaries. This can be enabled with incentivized systems as recommended in the biorights paradigm. Perusal of studies reveal that low greenhouse gas emission in tropical Nations having high forest cover, indicate the deployment of natural climate solutions therein, which can mitigate over 50% of national emissions mainly through avoided deforestation. In countries like India, Griscom et al. (2020) reports that strong governance and intermediate financing capacity increases the potential of nature-based solutions for mitigating greenhouse gas emissions. In another case where investment has been made on low diversity plantations, as because the policy is not grounded on biodiversity conservation and Restoration of ecosystems, it is found that fast growing monocultures sequester carbon rapidly but they cannot maximize carbon storage for a long term, as they are vulnerable to diseases pests and climate extremities. In contrast to this naturally regenerating forest, like mangroves, have high carbon sequestration rates as well they are rich in biodiversity and resilient to climate extremes, diseases and pest attack. Lewis et al. (2019) marked that any naturally regenerating forest can sequester 40 times more carbon than commercial plantations. Moreover, commercial plantations use non-native species and they may become invasive leading to introduction of new diseases, creating water scarcity and disturbing the local vegetation dynamics (Smith et al., 2017). Those ecosystems, which are prone to drought or fire and do not naturally thrive, there afforestation may reduce resilience to climate change and can compromise long-term carbon storage (Veldman et al., 2015).

Several diverse ecosystems can deliver much wider range of regulator and cultural ecosystem services which increases the cost effective of nature-based solutions. Biorights always tries to leverage the scope of intensifying such ecosystem services so that the payments thereof can be substantial to compensate opportunity cost of the beneficiaries. It is often found that policies that offer financial incentives for scaling up nature-based solutions towards reducing Greenhouse gas admission make compromise with the local land rights and this may lead to land grabs denying the rights of marginal communities. It is important to mention in this regard that Warsaw framework for REDD+ or specifically recommended the recognition of local community rights and respecting the socio-cultural aspects of indigenous communities as mentioned in the Cancun safeguard (UNFCCC, 2011). However, the operational guidelines for or implementation monitoring and evaluation remained very vague in all such recommendations. Another simple case study where introduction of invasive non-native plants had devastating impacts on the local biodiversity clearly explains that any nature-based solution must be grounded in robust understanding of the geographical distribution of the biomes, value of the local biodiversity and as well their ecological resilience. Thus, any nature-based solution for climate change mitigation does not curtail or distract the development objectives from decarbonizing the economy (Anderson et al., 2019). This is the unique space where biorights interventions create a Win-Win situation safeguarding the interest of local communities and conserving the habitat and the biodiversity in the locality. This entirely depends on the innovation partnership and monitoring of the intervention at its grassroots.

There has been a clear bias in promoting engineered solutions for climate change mitigation in the developing Global South. This is probably because of the economic drive that attracts direct Investments generate employment opportunities and boost the local economics. However, the impacts of engineered solutions and their sustainability has been under the lens for years. Many locally adopted nature-based solutions have been found to be more resilient sustainable and adaptive. Whenever any intervention is designed in alignment with the biorights model it is therefore important to understand three things of which the first one is the extent to which the solution will be exposed to climate change impacts in that particular place region or ecosystem. This also includes the exposure of the local communities in the intervention area. Secondly it is important to estimate the sensitivity of the local system to the impacts of climate change. For example, a coastal village, where the main livelihood activity is agriculture and fisheries, will be more sensitive to the impacts of sea level rise and salinity Ingress. Based on this the design of biorights intervention needs to be developed. Thirdly the adaptive capacity of the habitat and its people are also very important for the success of any such intervention. This provides both time and space to adjust or innovate in response to the changing conditions. Any biorights intervention act at the interface between social economic and ecological assistance to reduce the vulnerability of each of them and as well the system as a whole. Therefore, it always has a positive influence over all three dimensions of socioeconomic vulnerability and environmental crisis.

The first dimension of vulnerability framework is reducing socio economic on communities to the immediate impacts of climate change. Therefore, protecting restoring or managing natural ecosystems as an integral part of nature-based solution is very important. Examples like natural agro-farming practices and cropping among forest trees are found to be good for enhancing yields in variable climate regime. Similarly vertical gardens and green roof in urban settings can help in abating heat island effect as well contribute to clean ambient air (Alexandri & Jone, 2008; Bowler, 2010; Liu et al., 2014; Kabisch et al., 2016).

Socio-economic sensitivity in climate change is another dimension or of the vulnerability framework. For example, agro-pastoral communities in the degraded semi-arid rangelands of Kenya where found to be more resilient to climate change when they were rehabilitated and they used enclosures to keep their livestock. Capacity building in Alternative livelihood and diverse sources of income reduced their sensitivity thereby improving the standards of Living. In Zimbabwe the communities accomplished a degree of food security by protecting the forests and practicing Agroforestry. Nature-based solutions or ecosystem-based adaptations in interventions with bio-rights component can always enhance the delivery of ecosystem services that sustain livelihood and community wellbeing as well provide diverse sources of income to overcome economic crunch during climate-related hazards.

The third dimension of vulnerability framework is adaptive capacities in socioeconomic stature of the communities. Nature-based solutions are designed to support species richness and conserve the genetic pool in wilderness so that it can augment climate resilience. Such adaptive capacities also help in sustainable intensification of ecosystem services as well develop a robust local economy. Several other influential factors like financial inclusion, human resource development, education and governance also improve the adaptive capacities of communities. Biorights can very well provide the scope of developing adaptive capacities true ecosystem-based adaptations and improved nature services that can compensate the opportunity cost of poor communities. Similar examples can be cited from Togo and Sri Lanka for Bangladesh (Woroniecki, 2019; Munang et al., 2014).

All the three dimensions of vulnerability framework can affect the ability of an ecosystem to be e a sink for carbon dioxide emission that is meant for reducing climate vulnerability. This explains the effectiveness of nature-based solutions and its management approaches. However, with the increasing frequency and intensity of climate events and other stressors like land use change and pollution has become a hindrance for sustainable management of the solution. The typical approach of bio rights at community ecosystem interface helps to be resilient to these externalities. In many cases increasing frequency of climate led disturbances can result in a dramatic decline in the adaptive capacity of the ecosystem and the community as well. This will have an impact on the habitat and nature services thereby increasing the volume of Lost opportunities. For example, drastic changes in sea level can be e of much less harm to mangroves then the salt marshes. Therefore, the composition and tolerance of species as well as ecosystem respectively needs to be e kept in consideration while designing a biorights intervention. With or without active management many ecosystems are in the process of transitioning two alternative States under climate change idiot for example the domination of algal scum after mass Coral mortality in the reef of coastal habitat clearly explain that such transitions cannot support human adaptations. Therefore, the model of bio rights intervention must be design with the flexibility of provisions for adaptation. It is imperative to note that sometime communities can also so provide similar adaptation benefits in the pre disturbance period and may provide additional novel adaptation services following such disturbances. For example, in the delta region of south Bangladesh where most of the land get inundated with flood waters, the communities have developed a flood resilient float farming practice to enable primary productivity in the flooded area. However, when we design a biorights project, it is important to understand the performance of the project under varied climatic conditions, drawing on the knowledge of the eco-evolutionary mechanism that reinforces the ecosystems capacity to resist and recover from major disquiets through adaptive and mitigative pathways.

Over a decade several development agencies like UN environment, UNDP, Food and Agriculture Organisation (FAO) Birdlife and WWF advocating for nature-based solutions being led by communities at ecosystem level as a major approach for disaster risk reduction and sustainable solution. The reason behind such decisions goes to the fact that nature solutions are low investment resilient and sustainable solutions compared to engineer solutions. Additionally, ecosystem-based adaptations always deliver better nature services with a wider range, the for the benefit of the marginal communities. This can be a buy-back point for the bi rights projects. It is also important to understand the science behind the cost effective of nature-based alternatives so that we can find synergy among different solutions for or more guaranteed sustainability.

One of the major constraints in deploying ecosystem-based adaptations is the difficulty in measuring their social ecological effectiveness under the influence of fluctuating factors in the given context over space and time. At the same time the efficacy depends on the perspective of needs of the beneficiaries for whom the intervention has been designed. In many cases unexpected changes in community decisions social political interference or sudden changes in the ecosystem conditions main pose extremely challenging situations for measuring the effectiveness. At the same time few interventions improve their effectiveness only when they are subjected to climate disturbances or other similar perturbations. In Deltaic parts of Eastern India, nature based solutions for stabilizing mud embankment and preventing salt water Ingress, proved very efficient when the sustained two super cyclones in the year 2020–2021.

Similarly in assessing the cost effective of any ecosystem based adaptive solution, it becomes necessary to undertake social audits and other intangible benefits like intensification of ecosystem services or enhancement of carbon sequestration potential to draw the conclusive benefits to cost or returns on investments. Renewal repair and renovation is an essential factor for or extended durability of any intervention weather engineered or natural. Estimating the cost with a life cycle approach in these cases raisins another question of balancing future benefits with current cost. In many situations nature-based interventions may be natural Eyes increasing the resilience of the ecosystem where it becomes difficult to predict the benefits to cost as compared to other infrastructural solutions. All these bottlenecks prompt for a sustainable mix of grey and green solutions, where the opportunity cost can be directly compensated from the payments of ecosystem services, without compromising with the conservation objectives in the given ecosystem.

7.4 Why Should Biorights Work in South Asia? – Some Success Stories

(a) **Evaluating Prospects of BIORIGHTS: Successful Case studies from Asian Countries**

7.4.1 GPI and Poverty Reduction: Indonesia

At the 8th Conference of the Parties of the Ramsar Convention in Valencia in November 2002, during which the publication on Wise Use of Mires and Peatlands was launched, a draft Resolution [COP8 DR11] on designation of under-represented wetland types was signed, which included reference to the role of the Global Peatlands Initiative (GPI). Following this, in the stakeholder meetings in Central Kalimantan of Indonesia in relation to the Ex-Mega Rice Project, which affected over 1.5 million ha of tropical peat swamp forests and caused significant environmental problems and increased poverty as well, the GPI financed the development of the "biorights" financial system, which provides a basis for addressing poverty-environment issues. During the UNEP-Finance Roundtable in Tokyo in October 2003 in a workshop on Biodiversity and Finance, the innovative biorights approach was presented by the authorities of Central Kalimantan and representatives of the local *Dayak* communities. Other GPI Projects in Indonesia (Sumatra) focused on community-based peat fire prevention, including capacity building, awareness drive and extinguishing fires in critical areas as well. The peat fires in South-east Asia have a huge impact on the local economy as a result of resource destruction, as well as impact on transport, health and life. The GPI projects were some of the first to actually set examples on how this major problem can be addressed through bottom-up approaches involving local communities and other stakeholders.

7.4.2 PES: An IUCN Endeavor in Vietnam

The loss of Vietnam's mangrove forests is a complex issue. To blame the farmers who are clearing mangroves whilst ignoring government policy and an expanding aquaculture industry is to blame the victims. In the context of ever-increasing shrimp exports, simply moving villagers and planting mangrove trees is unlikely to solve the problem. IUCN, through the Wetlands Education Network of Viet Nam (WENVN), partnered with the Can Gio protection forest management board to organize the meeting in Can Gio District, Ho Chi Minh City, to identify options about Payments for Environmental Services (PES) system for the nation's wetlands, and specifically for Can Gio's mangroves. PES is a resource management tactic that aims to put the environmental benefits of a given ecosystem on par with other remunerative uses by giving those services a market value. It requires the beneficiaries of a given ecosystem, who receive services like clean water, fisheries, disaster mitigation or ecotourism opportunities, to compensate local communities and others who are responsible for maintaining ecosystems. As such, it acknowledges the role people play in protecting ecosystems and provides them with an incentive to do so, rather than exploiting the environment for short-term economic gain. IUCN helped pioneer PES in Viet Nam, by assisting the government in drafting the Law on Biodiversity containing a related article. The WENVN contributes to the management, conservation and sustainable use of wetlands in Viet Nam through education, awareness and communications activities. It does so partly by creating a forum where organizations can share information on wetlands conservation, this time about PES.

The consultation meeting of IUCN introduced the concept of PES to stakeholders in Can Gio. About 40 participants from government agencies, academic institutions, tour companies and conservation groups brainstormed and evaluated the services the local mangrove forests provide, and discussed who benefits from those services, as well as which groups are in the best position to protect them. Participants said they expect the findings from the consultative workshop to contribute considerably to efforts to identify options and craft proposals for suitable and effective PES mechanisms for wetlands in Viet Nam. More recently PES system with community partnership has shared the challenge of conserving the mangroves.

7.4.3 Community Forestry Initiatives: BIORIGHTS in Philippines

The reforestation initiative of locals in Philippines were responsible for the vigorous growth of a robust young mangrove forest through wise-use and adaptive management. Their lively hood has been very much associated with the sea and was included in the category of fisher folks. The edible organisms that thrive successfully in this mangrove forests are species of shelled mollusks and the crab. The ecological parameters such as prevailing ranges of temperature, salinity and all other physical and biological components seem to suit the growth and development of the edible crab and mollusks leading to their continuous production and harvest. These serve as food for the local community especially during emergency. The community was also able to establish a netted pond and allowed commercial brackish water fishes to grow and proliferate. In exchange for maintaining the mangrove, the caretakers were given crabs, mollusks and also fishes by the community manager. Though they do not belong to the poorest of the poor but it is exhilarating to know that they are earning money and at the same time helping in the maintenance of the ecosystem of the reforested mangrove making it ecologically healthy and economically sustainable. The municipal government and an NGO are behind the initial implementation and their continuous support as well as the proper maintenance of the will soon make this mangrove area not only as a food depot for humans but as a pioneering model for economic sustainable development and healthy ecosystem rolled into one.

(b) **India's First BIORIGHTS Project in East Kolkata Wetlands – A Threatened Ramsar Site**

7.4.4 East Kolkata Wetlands: Ecological Profile

Sometimes called the "kidneys of Kolkata (Calcutta)," the East Kolkata Wetlands (EKWs) are the largest resource recovery system of their kind in the world that recycles 780 million litres of raw sewerage per day. It covers an area of 12,500 hectares which were designated as a Ramsar site in August 2002, owing to its unique multiple-use resource recovery system sustained by the fishermen community. The wetland lies east of the city and includes a garbage dump known as Dhapa Square Mile, a mosaic of vegetable fields, a series of 264 odd fishponds connected by major and secondary canals, rice fields and vegetable gardens and 43 villages with 120 thousand people in all. The EKWs extend almost equally on both sides of a Dry Weather Flow Channel, which discharges into the Kulti Gang (the wastewater outfall of Kolkata 28 square kilometers to the east). The area is divided into eleven zones and includes four major sub-regions: freshwater fishponds, brackish fishponds (both of which are known as *bheris),* garbage farms, and paddy lands. Some thirty kilometers eastward, the river Kulti-Bidyadhari drains into the Bay of Bengal. Over the surface the fishponds produce some 13,000 tonnes of fish annually, whose yield are 2–4 times higher than average fish ponds, is among the best of any freshwater pisciculture in the country. 11,000 tonnes of rice per annum and some 150 tonnes of vegetables per day are harvested from small-scale plots irrigated with wastewater. The vulnerable ecosystem harbours 206 species of wetland flora and fauna that includes the Red Listed Marsh Mongoose and Mud Turtle, 52 endemic species of fish and nearly 20 mammals. 40 migratory birds have been reported to visit these wetlands in recent past.

7.4.5 Economics & Land Ownership

There are various types of land ownership of fish ponds in EKWs. Most of them are owner-managed. Some 27 Cooperatives with about 2900 fishermen and women, operatesand functions well without any external assistance. Only six cooperatives are under the direct control of West Bengal State Government run Fisheries Corporation. Much of the land is also leased out by absentee landowners, some of which are further sublet. The opportunity cost and economics of waste water production has not been studied extensively, and the reliability of existing data was poor, as private farmers were not willing to disclose and the State-managed fishponds were too few to fully represent the entire area. Despite their potential to provide multiple natural and social services, the wetlands are an informal system of aquaculture. 'Community-ecosystem' approach under the aegis of biorights project led by South Asian Forum for Environment [SAFE] and supported by the Government of West Bengal, the of study on transformation of nature services to alternative economic opportunities in various peri-urban wetlands of East Kolkata enabled all stakeholders to partner equitably and allow rectification through adaptive management practices over time and space.

7.4.6 Urban Development and Encroachment Issue

The eastward expansion of the metropolis of Kolkata, encroaching the EKWs designated area is a threat to the future of the wetlands as it makes locals wary of investing their labour in long-term maintenance and upkeep the unique mosaic of the waterbodies, fishponds and marshes. Some unregulated industries, such as tanneries, have been releasing untreated effluent directly into the wetlands, further threatening the water quality. Siltation of the ponds is due to construction of concrete structure, marble factories and also through natural processes, the build up of silt has reduced the capacity of fish ponds over time. Some of the ponds those were initially six feet deep, have silted up to one and a half feet from the surface, reducing the fish production by two-thirds. The dredging and transport needed to clean out the larger ponds requires a large investment, even though the silt can be used for landfills planned for new townships north of the city. As this is an urban area, there are more competing, conflicting interests and agencies than would be found in a village. Moreover, unlike rural development, the issues are less black-and-white, for example, income opportunities exist locally whether or not the wetlands are preserved. However, the income and quality of life for those who depend on the wetlands might change dramatically if they were to disappear.

7.4.7 *The Biorights Project at East Kolkata Wetlands*

The first biorights project in India, launched at East Kolkata Wetlands on 2nd February World Wetland Day 2007, by South Asian Forum for Environment (SAFE), a regional civil society organization working towards sustainable environmental development through community based participatory programmes and partnership in Indian eco-region. SAFE had proposed to design a concept on biorights project that can assure sustainable fish production and biodiversity conservation at EKW along with a task to alleviate poverty in East Kolkata Wetlands. The pilot project intended to transform nature-services like ecotourism and handcrafts from water hyacinth weeds of wetlands as a financial tool for poverty alleviation and inclusive development of communities in these peri-urban wetlands through community participation and partnership. The specific objectives of the project that was funded by Innovative Challenge Fund, of Govt. of West Bengal and DFID (UK), has been:

- To identify economically beneficial nature-services in the area for compensating the opportunity costs of the communities
- To expand economic opportunities through capacity building among local beneficiaries, especially women, for livelihood
- To conserve the habitat and biodiversity of the wetlands through stakeholder partnership and community participation
- To create awareness among local stakeholders about restoration of East Kolkata Wetlands

7.4.8 *Community Involvement*

The community dwellers and stakeholders of EKWs were engaged through participatory decision making and equitable partnership in planning the project, as an integrated approach for sustainability. While capacities were built in alternative economic opportunities for compensating their opportunity costs in ensuring conservation of the wetlands, capacities in wetland stewardship were also built for the needful in the community members. The whole community shared the responsibility of periodically validating and terminally evaluating the impacts of the biorights scheme and thus they ensured adaptive project management.

7.4.9 *Sustainable Replication*

Replicability requires an integrated approach to orchestrate the synergies and the trade-offs between economic returns, societal equity and conservation objectives that comprises the "bottom lines" of sustainable development. Strategic Impact Assessment functions as an yardstick for assessing sustainability mechanisms and suggesting further adaptive management across space and time. The project could therefore be successfully scaled up in DeeporBeel Ramsar wetland in Assam and as well in Loktak Lake Manipur in similar framework. That it has a low investment and high return portfolio assuring endeavour it has a common acceptability feature. The project successfully testified the potentials of "Bio-Rights" in risk coverage through group microinsurance schemes, which endorsed better adoption of the project by the community dwellers.

7.4.10 The Partnership Initiatives

The periurban community accepted this as the first major endeavor in East Kolkata Wetlands that takes care of their livelihood through spreading the financial risk of fisheries by recovering the loss with another economic opportunity, as well extend risk coverage through insurance.

As an impact of extensive awareness building, the urban response on biorights was equally good. A general consensus in 'willingness to pay' approximately 18–20% higher value than normal tourism in the wetlands could be utilized for micro-insurance coverage of the wetlanders with a premium as low as USD 0.10 per month per family against health and disaster related coverage.

The formation of Self-Help Groups that SAFE could establish will be further strengthened with micro entrepreneurships and credit linkages. The fisherwomen will be exposed to hospitality training and young men will be trained in marketing skills. In the days ahead, micro financing for these groups, which is being assured from nationalized banks like SBI and Indian Bank, will ensure more economic stability and empowerment. It is expected that all such endeavors will help in bringing attitude change and raise the livelihood status of the wetlanders.

7.4.11 Impacts and Outcomes

This ecosystem at EKWs shows the significance of reuse in waste disposal and sanitary engineering which has not been well appreciated otherwise. It's been indicated that integrated resource recovery and institutionalization of urban solid and liquid waste management can reduce costs to municipalities by 30–90%. If the wetlands fall prey to urban expansion, the results will be not only the loss of food and livelihood, but will require the construction of an expensive sewage treatment facility to replace the services it currently offers. This case demonstrates the value of "invisible services" such as waste management, food production and livelihood, and it highlights how nature provides these services on a relatively large scale and in an urban environment. However, the integral part of this system is the wetlanders who through a community-ecosystem partnership ensure the sustainable development. But owing to urban encroachment their existences are at stake, though the wetlands are more significant to the urban growth and development issues. In the given situation the concept of BIORIGHTS has not only assured the alternative economic opportunity to the community but has also accentuated the urban partnership through the compensation points. The overall impacts of the project till date is briefly outlined herewith.

1. Nearly 35 fisher's cooperatives are self-sustaining this paradigm for last one decade, since the inception of the biorights project, wherein more than 5000 fishers are actively participating in the program, keeping the services of the wetlands alive and the habitat rejuvenated. It has a span of almost 3500 hectares of wetlands where sustainable aquafarming is being practiced.
2. Nearly 750–800 women wastewater fishers and informal women workers have formed a micro enterprise (http://resolvetrash2cash.org/) for recycling wastes that damage the wetlands and as well manufacturing products of water hyacinth fibre that grows as an exotic aquatic weed in the wetland, increasing eutrophication and impairment of ecosystem services.
3. There has been a robust impact on policy implication, thereby saving the wetlands from getting de-listed from Ramsar's International tag, as well getting extended support from National Bank for Agriculture and Rural Development (NABARD) for endemic local fish farming and micro-financial help to the women self help groups. Currently the State Government in collaboration with the Wetlands International is developing a community-based sustainable conservation plan for the entire stretch of wetlands.
4. The intervention has been acknowledged globally through reports and accolades from Ramsar Secretariat, Nairobi Work Program of UNFCCC, The Economics of Ecosystem and Biodiversity, UN Environment (http://teebweb.org) etc. Institute of Global Environmental Strategy (IGES), Japan has made a study on the emission abatement potentials of the recycling unit run by the women therein (https://www.oneplanetnetwork.org/sites/default/files/sle_project_synthesis_story_200901.pdf).

(c) **Integrated Mangrove Aquafarming in Deltaic Sundarbans in 24Pargana District, West Bengal**

7.4.12 Shrimp Farming and Mangrove Loss

The mangrove ecosystems are one of the world's most diverse environments, providing a variety of ecosystem services critical to human well-being. These conservation efforts are estimated to be worth 14,000–16,000 USD per hectare per year, totaling 1.6 billion USD per year for all mangrove forests worldwide (Barbier et al., 2011). One of the most essential functions provided by these ecosystems (especially in light of current global climate change) is their capacity as carbon sinks. Mangroves have the ability to collect and store carbon from the atmosphere, which is then embedded in coastal sediments as 'blue carbon'. The capacity of mangroves to store carbon and mitigate greenhouse gas emissions became prominent at the beginning of the twenty-first century (Donato et al., 2011). Since then, mangroves have gained international interest for their potential to contribute to carbon mitigation strategies and for their ecosystem services that support adaptation to climate change (Lovelock & Duarte, 2019). However, pollution, deforestation, fragmentation, rising sea levels, and, most importantly, land use changes pose a threat to these ecosystems. Other purposes encompass fish farming ponds, along with urbanism, agriculture, and infrastructural facilities (Arshad et al., 2018; Ragavan et al., 2019).

The growth of mangrove species can be better in lower soil bulk density because for its capacity to provide organic matter and protect the coastal areas from erosion processes. The presence of roots and litter that serves as habitat of the wild aquatic biota, and its consequent activity, result in an increase in sediment pore and macropore development. Mangrove conversion to shrimp farms result soil compaction, thus increased bulk density and decreased soil porosity. The soil organic carbon (SOC) content decreases due to conversion of mangrove to shrimp farm, SOC in mangrove areas is likely a result of litter falls and underground roots instead, in the shrimp farm, the vegetal cover is scarce. The decline in the amount of organic matter in the shrimp farms has been attributed to decomposition of the organic matter in soils especially during periods when the ponds are drained after shrimp harvests. Draining and limiting tidal flows create aerobic conditions favorable for carbon oxidation. Shrimp farms shows decline in the carbon stocks compared to mangroves. Soils are the almost only carbon pool for shrimp farms where the vegetative cover overlaying ponds has disappear and carbon stock is limited to that contained in soil (Arifanti et al., 2019). The average potential CO_2 emission associated with the transformation in land use of Brazil has been established at 139 kg CO_2e m^{-2} (Kauffman et al., 2018) and in areas of Indonesia at 226 kg CO_2 m^{-2} (Arifanti et al., 2019). Globally, shrimp farming expanded rapidly in the 1980s and 1990s, mostly in the tropical and subtropical regions of Asia and South America, driven by growing international demand and a high market price (Primavera, 2006; FAO 2007; Bush et al., 2010). As a consequence, huge expanses of mangrove habitat were converted into intensely managed aquaculture systems, which had higher carbon footprint and generated high levels of pollution from fertilizers, pesticides and antibiotics. Vaiphasa et al. (2007) reported important adverse effects of the shrimp pond effluents on the survival of mangroves surrounding them.

An estimated 1.5 million ha of mangroves were lost in the world due to shrimp culture since the 1980s with a global loss of blue carbon stock of about 831 million t (Ahmed et al., 2017). Although most countries have recently restricted the conversion of mangroves to aquaculture and other development activities, the destruction of mangroves continues in the twenty-first century. Overall, 336, 300 ha (2.1%) of global mangrove forests have been lost from 2000 to 2016 at an average annual rate of 0.13% and commodities driven land-use changes such as aquaculture and agriculture accounted for 62% of the total global mangrove loss (Goldberg et al., 2020). The deforestation rates were very prominent in the Bay of Bengal marine province where 74% of mangrove land was converted for agriculture and aquaculture during the last 16 years. It is predicted that emissions from mangrove loss could reach 2391

Tg CO_2eq by the end of the century and the highest emissions were predicted in the southeast and south Asia (West Coral Triangle, Sunda Shelf, and the Bay of Bengal) due to conversion to aquaculture or agriculture (Adame et al., 2021).

Currently, 82 countries and regions are reported to produce farmed shrimp covering all the continents although 85% of shrimp production is concentrated in Asia. In 2018, global farmed shrimp production hit a new high of 6 million tonnes, accounting for 7.3% of total farmed aquatic animal production. The giant tiger prawn (*Penaeus monodon*) used to be the dominant species of farmed shrimp contributing 51.0% of the world production in 1998 and its contribution to world production had dropped to 12.5% in 2018. There are two farmed shrimp species in India: *L. vannamei*, also known as white leg shrimp, and *Penaeus monodon*, or *P. monodon* (black tiger shrimp). India currently occupies a competitive position in the global shrimp supply chain and is the fourth-largest shrimp producer in the world, with a global market share of 11.4%(Miao & Wang, 2020). Amid the COVID crisis, the country produced 8,43,361 metric tons of shrimp in 2020 and has grown at a 22% Compound Annual Growth Rate (CAGR) since 2010, which is higher than the global average. Indian shrimp farming sector has shifted its focus from native *Penaeus monodon* to exotic *L. vannamei* and currently, it occupies 97% of total shrimp production in India growing at a 53% CAGR over the last ten years (2010–2020). Simultaneously, *P. monodon* production showed a decreasing trend with a 15% decline rate over the last ten years.

In India, mangroves are found along the coastlines of nine States and four Union Territories (UTs) and are spread over an area of 4975 km^2, which represents 3.6% of global mangrove vegetation and 0.15% of the total geographical area of India (FSI, 2019). 21,268 ha of total mangroves and mudflat habitat had been imperiled for shrimp farming between 1988 and 2013 in the major shrimp producing states like West Bengal, Andhra Pradesh and Odisha (Jayanthi et al., 2018).India's shrimp industry is flourishing, but it faces many threats and sustainability challenges. Increased disease-associated production loss due to environmental degradation through mangrove deforestation is one among the other threats. Furthermore, limited value-added processing, uncertainties in the market and lack of traceability regulations are aggravating this situation. Climatic variability is also threatening the viability of the shrimp farming system in both direct and indirect ways.

7.4.13 Integrated Mangrove Aquaculture: A Potential Nature-Based Solutions

Integrated mangrove-shrimp cultivation or simply integrated mangrove aquaculture (IMA) is the coexistence of mangroves and shrimp aquaculture in a tide-fed environment and has been recognized as environmentally friendly low-input farming system. Commonly known as silvo-aquaculture or silvo-fisheries, it is best suited in extensive to modified extensive shrimp farming system and traditionally it is practiced in many Southeast Asian countries like Indonesia, Philippines, Thailand, Vietnam, Malaysia, Cambodia and Myanmar. These have a lower capital requirement compared to other shrimp farming systems, livelihood diversification through polyculture, provision of regular income and recognition as an organic farming practice. In the integrated system, mangrove integration can be in 30–70% of the active farm area, the shrimps are raised in the ditches between the rows of planted mangrove on platforms while large areas of water are intermixed with a wide mangrove area in the associated system. Shrimp ponds and mangrove patch are clearly distinguished by dykes in the separate system, where mangrove chunk is used as biofilter for the farm effluent (Primavera et al., 2007; Bosma et al., 2014). Stocking densities of shrimp post-larvae varies and ranged <10 to 10–20 PL m^2 year ^{1}. Part of the shrimp yield of the mixed mangrove-shrimp farms depends on the natural recruitment of production stock. Shrimp yield is higher in ponds with 30–50% mangrove cover and strands younger than 5–10 years with timber production about 10 m^3 ha^{-1} $year^{-1}$(Bosma et al., 2014; Tran, 2005). The annual productivity of shrimp varies greatly in the IMA system but remained <400 kg ha^{-1} $year^{-1}$ while productivity ranges from 700 to 800 kg ha^{-1} $year^{-1}$ in case of the fish-shrimp polyculture system (Bosma et al., 2014).

Mangrove species should be chosen carefully for a new IMA system because shrimp productivity is influenced by the inundation frequency, tannin content, half-life and decomposition rate of mangrove leaves. The leaf litters of different mangrove species contain different toxic or beneficial constituents. The tannic acid from tree leaves is potentially toxic to aquatic organisms while increasing leaf litter concentrations increased Ammoniacal nitrogen (NH_3-N) concentration due to organic matter degradation if water exchange did not do properly (Rejeki et al., 2020). *Rhizophora* and *Avicennia* sp. were frequently planted with low risk of aquatic production (Primavera, 2000). Leaves of *Avicennia* increased pond fertility and regulated pH during the rainy season while leaves of *Rhizophora* provide a suitable substrate for the growth of favorable periphytic beneficial biofilms. Leaves of *Ceriops*, *Rhizophora* and *Sonneratia* have antiviral effects that can mitigate WSSV in shrimp ponds (Bosma et al., 2014).

7.4.14 Indian Scenarios of Integrated Mangrove-Shrimp Cultivation

M. S. Swaminathan Research Foundation (MSSRF) is pioneer in introducing this approach named Integrated Mangrove Fishery Farming System (IMFFS) in India with an aim to rehabilitate abandoned shrimp ponds and to restore traditional mangrove-based livelihoods. Since 2006, MSSRF had been successfully tested the IMFFS model in different coastal hamlets of Tamil Nadu where shrimp (*Penaeus monodon*), crab (*Scylla* sp.), and fish (*Mugil cephalus*, *Lates calcarifer*) are farmed. In the IMFFS system, shrimp ponds were modified through outer and inner embankments on which certain mangrove species (*Rhizophora mucronata, Rhizophoraapiculata, Avicennia marina*) and other saltmarshes (*Sesuviumportulacastrum*) were planted. It is important to note that the survival and production of fish especially *Lates calcarifer* from IMFFS is higher by 11% and 12.5% than the open aquaculture system.

7.4.15 Traditional Shrimp Farming in West Bengal

West Bengal on India's east coast is one of the major shrimps producing states and holds the largest farming area nearly 50,000 ha. The coastal region of West Bengal is positioned in the tidally active lower deltaic plain of the Ganges-Brahmaputra-Meghna basin and hosts the largest contiguous mangrove forest ecosystem in the world— the Sundarbans. With tidal inundations from Bay of Bengal water twice a day, the region offers mangrove-based nutrient-rich saline water through its estuarine networks, plenty of wild finfish and shellfish seeds and acts as a bio-shield from natural and climatic hazards that provides ample space for brackish water aquaculture in the tidal dominated areas of West Bengal. Once mangroves dominated, reclamations of mangroves from these areas began in 1770 during the British colonial period and continue until the mid-nineteenth century primarily due to agriculture and other land revenues. Since the 1970s, the region witnessed a rapid transformation from the paddy cum fishery integrated system to a traditional brackish water shrimp-fish mixed farming system. The rapid growth of shrimp farming simultaneously rises many social conflicts like salinization of soil, conversion of agricultural land and mangrove destruction. The shrimp farming situation is further aggravated in 1994–1996 and early 2000s due to the mass outbreak of White spot disease caused by white spot syndrome virus (WSSV) resulted in mass mortality of cultured shrimp and shrunken shrimp aquaculture industry. The export production of *P. monodon* from West Bengal has been shown an 84% upsurge during the past decade with an increase from 26,800 t in 2001 to 49,319 t in 2017. The shrimp farming sector of West Bengal was massively affected in the post-Covid scenario due to the disruption of supply chains and its contribution to the total shrimp production lowered down to 7% in 2020. Surprisingly *P. monodon* production in West Bengal witnessed a declining trend since 2018with −5% CAGR in the last 10 years (2010–2020), while *L. vannamei* production has increased manifold with a CAGR of 61% in the last eight years (2013–2020).

7.4.16 Ecosystem Based Production Methods

Parts of North and South 24 Parganas districts of West Bengal sustain traditional low-laying tide fed aquatic food production ecosystem which is unique in South Asia. These traditional aquatic water bodies are very big ranging from 0.5 ha to

100 ha bordered by embankments and shallow typically less than 1 m depth with characteristically very low stocking density, locally called 'Bheri'. *P. monodon* is the main target species due to its attractive body colour, unique taste and expanding demand in the urban and export market. The medium (15–20 ppt) and low saline (0–12 ppt) *bheries* are prevalent in North 24 Parganas occupies 40% of waterbodies in the district. These *Bheris* provide unique ecosystem services like blue carbon sequestration, flood control, supports biodiversity, etc. and have immense social and economic values with traditional knowledge and wisdom. Bidhyadhari river is the lifeline of shrimp farming in the North 24 Parganas which also serves as an outlet for sewage and storm water from Kolkata and North 24 Parganas. Bidyadhari receives saline water under tidal effect from the sea and is discharged into the Bay of Bengal through the Haribhanga estuary. The water quality of this river is rapidly deteriorating due to metropolitan sewage and effluents released from various factories, including tanneries, garment, plastic and glass factories.

In this system, tidal brackish water from the adjacent creeks or man-made canals is let into the farm by gravity flow and entry of water is normally regulated through a wooden sluice gate made of planks and bamboo poles (locally called *Goi*). Pens made of bamboo screens locally called *Pata* are used for filtration of the water. Seeds of different fish and shrimp species also find entry into these screens along with the tidal water (auto stocking). Stocking and harvesting are periodic which are according to the lunar phase. The stocking is generally starting from late January and continues up to mid-October. Partial water exchange is done during the lunar phases taking advantage of the tidal amplitude of spring and neap tides. In most of the *Bheries*, a single sluice is used for both letting in and draining out the water. Usually, no to limited supplementary feed is used, and shrimps grow based on periphyton-based natural productivity. The natural growth of filamentous algae, *hydrilla*, etc. in these large and shallow farming areas provides an excellent environment for the growth of tiger shrimp. Harvesting is normally done during the full moon and new moon period using traps made of bamboo locally called *Atol* and partially by cast netting. Harvesting starts 3–4 days before the full moon and new moon and continues 3–4 days after the full moon and new moon. Complete harvest ended in the second week of December. In this traditional shrimp-fish polyculture system, salinity fluctuates throughout the year and gets its peak in the summer season but remained at 8–10 ppt. Among the wild stocks, Horina shrimp *Metapenaeus Monoceros*, Indian white shrimp *Fenneropenaeus indicus*, sea bass and mullets are the major ones. Apart from *P. monodon* seeds, farmers also stock freshwater prawn *Macrobrachiumrosenbergii* as the species grows well in mild brackish water conditions and fetches good market price. Tilapia (*Oreochromis niloticus*) is another preferred species due to its faster growth rate and salinity resistance. Brackish water finfish species like *Lates calcarifer*, *Chelonparsia*, *Mugil cephalus*, *Mystusgulio* etc. are also stocked. Farmers also preferred to stock Indian major carps (IMCs) such as *Labeorohita*, *Catlacatla* and *Cirrhinusmrigala* as well as silver carp and grass carp when salinity becomes less. 80–90% of shrimp production of these *Bheris* are export oriented. Annual shrimp production from the *Bheris* is typically low and ranged between 300–400 kg ha^{-1} $year^{-1}$. Together with fish and other species, the total production was observed as 600–650 kg ha^{-1} $year^{-1}$. It is interesting to note that, after harvesting and dewatering of farms, majorities of the farms produce a substantial volume of freshwater Eel (*Anguilla bengalensis*) which is mainly caught by local indigenous people mainly tribals by handpicking method.

7.4.17 Challenges of Traditional Shrimp Farming

Poor quality of shrimp seed, irregular supply from hatchery, low survival rate and increase disease risks are the key challenges traditional shrimp farming system in West Bengal. Since couple of years, Bheri farmers are witnessing a decreased productivity of shrimp and poor water quality. It has been argued that wild shrimps are one among the causative agents of the WSSV mass outbreak flowed by industrial pollution in the source water. Amidst several adversities, still traditional shrimp farmers prefer to stock seeds from wild source as they firmly believe that growth, body coloration and survival rates of wild shrimps are far better than the hatchery-reared seeds. Some farmers experienced that salinity of the water in the North 24 Parganas areas are decreasing while it is increasing in Sundarbans area. Increased sedimentation in the navigating channel, turbid water, foaming of water, presence of toxic compounds creating several hardships in the shrimp yield at present context. Moreover, these kinds of negative inputs are adversely impacting the established export market. Furthermore, climate change induced extreme weather events, coastal flooding etc. have negatively influenced shrimp production in those areas.

7.4.18 Integrated Mangrove Aquaculture in North 24 Parganas district, West Bengal: A pilot initiative

Understanding the opportunities and potential benefits of mangrove integrations in existing traditional shrimp farming system, an initiative was started in 2018 involving local farming community. Subsequently Three IMA systems were developed in Basirhat-II and Haroa Blocks in North 24 Parganas district of West Bengal. The principal objectives of this unique nature-based solution were:

(i) Increase in shrimp productivity through mangrove-based nutrient recycling
(ii) Reduction of blue carbon emissions and biodiversity improvement
(iii) Improvement of overall shrimp farming environment and establish mangrove based organic shrimp farming system
(iv) Establishment of biorights in the envelope of environmental conservation

In a public-private partnership mode, the pilot initiatives were supported by the German retail chain 'ALDI SÜD' in cooperation with the European shrimp importer 'Shore' and German Development Bank KfW-DEG. The project was implemented by Germany based consulting firm *Bluesensus* along with its Indian partners (BSA & NEWS), institutions and local community. All farms are located within the organic shrimp farming clusters developed since 2014. Currently BSA has developed three organic shrimp farming clusters of 1532 ha involving 31 farmers with guaranteed traceability and chain of custody. The farms were certified in accordance with European Union organic regulation and Naturland standards since 2015. In 2018, after introduction of organic aquaculture standards under National Programme for Organic Production (NPOP) in India, the BSA project was pioneer in NPOP certification (India Organic) in West Bengal.

7.4.19 IMA Methods and Practices

All these three pilots are modified extensive shrimp farming system with strict adherence of management practices according to internal control system (ICS) and standard operating protocol (SOP) developed as per certification standards. All farmers stocked organic shrimp PL (<10 PL m^2) and only liming were allowed. Usually, no supplementary feed was given, and shrimp grew based on natural productivity of the farm. The farming communities (farm owners, farm workers, part-time workers) were empowered through different training and exposure visits that created a consensus about mangrove conservation among local communities. Details about three different IMA pilot locations was given in Table 7.1. Before planting mangroves into the farms, dykes were redesigned with different slopes to make plantation easier. Based on the size and topography of the farms, island like pits were also developed for planting of mangroves. Mangroves were incorporated into the inner embankments, peripheries, pit, and outer embankments of adjacent canal. Mangrove species were selected based on salinity and soil profile and varieties of freshwater-loving to salt tolerant mangrove species were tested in this intervention. Based on the seasonal availability, mangroves were planted in the beginning and end of the monsoon season. The species were collected from mangrove nursery located at Sundarbans. To provide some immediate benefits to the farmers, some fruit trees were also given which were planted in the periphery to the rest house. Protection through bamboo fences were ensured before planting to avoid livestock grazing. Regular monitoring was done to see the survival and growth of mangrove species.

7.4.20 Observations and Major Lessons Learned

After the subsequent three years of intervention, it has been observed that the pond water can be managed independently, and the shrimp culture is not affected by the mangrove. Shrimp and fish production of these three farms varied according to year and situations aggravated during Covid pandemic. It is very important to note that

Table 7.1 Different IMA systems in North 24 Parganas, West Bengal

Parameters	IMA-1	IMA-2	IMA-3
Implementing year	2018	2019	2020
Block	Basirhat-II	Haora	Basirhat-II
Village	Andulpota	Bireswar-Gopalpur	Sadiknagar
Geocoordinates	22°37′25.45″N 88°48′21.51″E	22°36′15.58″N 88°43′49.89″E	22°40′11.13″N 88°44′55.53″E
Area (ha)	5.89	1.45	2.0
Maximum salinity (ppt)	10–12	10–12	10–12
Water pH	7.5–8.5	7.5–8.5	7.5–8.5
Mangrove species introduced/ tested	*Avicennia officinalis* (Bain), *Bruguieragymnorrhiza* (Kankra) *Rhizophora apiculata* (Garjan), *Ceriopsdecandra* (Goran), *Heritiera fomes* (Sundari), *Sonneratia apetala* (Keora), *Nypafruticans* (Golpata) *Xylocarpusmoluccensis* (Pasur)	*Avicennia officinalis* (Bain), *Bruguieragymnorrhiza* (Kankra) *Rhizophora apiculata* (Garjan), *Ceriopsdecandra* (Goran), *Heritiera fomes* (Sundari), *Sonneratia apetala* (Keora)	*Avicennia officinalis* (Bain), *Avicenniamarina* (Peara Bain), *Sonneratia apetala* (Keora), *Bruguieragymnorrhiza* (Kankra), *Rhizophora apiculata* (Garjan), *Nypafruticans* (Golpata) *Heritiera fomes* (Sundari), *Xylocarpusmoluccensis* (Pasur)
Type of planting	Dyke, pit, peripheral planting	Dyke, pit, peripheral planting, outer canal bank planting	Dyke, pit, peripheral planting
Mangroves propagated in different intervals	2510	1576	3118
Mangrove coverage	10%	20%	15%

to understand and interpret any quantifiable impacts of mangrove integration into shrimp ponds, minimum 5 years will be required after planting as it directly linked with age of mangrove strands. Although, *Bruguieragymnorrhiza* (Kankra), *Rhizophora apiculata* (Garjan) are growing well in low saline areas across the three sites, growth of *Avicennia officinalis* (Bain), *Sonneratia apetala* (Keora), *Heritiera fomes* (Sundari) and *Nypafruticans* (Golpata) are remarkable. However, *Ceriopsdecandra* (Goran) showed poor growth rate in farm conditions. It is interesting to note that, flowers already observed in *Avicennia officinalis* and *Sonneratia apetala* within two years of planting. Livestock grazing is major threats in this kind of system as the green dykes of farms are openly exposed for grazing. After protecting the farms from livestock grazing through fencing, natural recruitment of mangroves was observed in two farms. In that case, fastest growth rates were observed for naturally settled *Sonneratiacaseolaris* (Ora), *Avicennia officinalis* (Bain) and *Excoecariaagallocha* (Genwa). Furthermore, to compensate mortality of mangroves, continuous supply of mangrove seedlings, propagules etc., from nurseries are important and was done in consecutive years after afforestation. Local women and tribal women are trained for preparation of mangrove nursery for continuous recruitment of mangroves. Capacity building also done for doing kitchen garden to meet the daily need of nourishment. Regular monitoring of plant health is essential in IMA system at least two years from plantation. Therefore, proper training and community participation are the crucial factors of success of the IMA venture. It has been observed that watering of plants during December to February as well as in summer seasons reduced mortality of mangroves. Furthermore, use of organic amendments enhanced mangrove growth.

7.4.21 Way Forward

Integrated mangrove-shrimp cultivation can be seen as tool for nature-based solutions to alleviate negative environmental impacts. A well-designed IMA system can offer a wide range of ecosystem goods and services that support climate change adaptation and mitigation. It is also a mangrove conservation and restoration strategies to compensate mangrove losses by shrimp cultivation and to sequester and store blue carbon. It also acts as biofilter for excess nutrient load and maintains water qualities of adjacent shrimp farms through nutrient exchange. Decomposition of certain mangrove leaves temporarily boosted shrimp growth and control diseases. However, important ecosystem functions of mangroves depend on management factors of the farms and its connectivity with the natural ecosystem.Mangrove rehabilitation in abandoned fishponds is often constrained by complexities in coastal zone management, land tenure and user rights. Increasing the economic returns by capitalizing on the ecosystem services through carbon credits and through organic certification with a premium price for the products, requires that the farmers organize in clusters or cooperatives, because the cost of individual registration and verification are too high. Successful mangrove restoration depends on various biological, hydrological, geological and environmental factors and synergy between ecological aspects and social aspects is the most essential element to achieve success. Involvements of local communities are critical in case of successful mangrove management. Interested shrimp farmers can be trained and guided towards responsible mangrove management so they can stimulate other farmers about potential of mangroves in augmenting shrimp productivity. For successful implementation of IMA system, however, considerable institutional support is necessary in the form of technical and financial assistance. For these initiatives to be implemented, scaled, and sustained, key stakeholders such as international organizations, researchers, policymakers, government and non-governmental organizations, and fish farming communities must engage together. Nevertheless, more comprehensive research is required to foster greater understanding on the impacts of mangrove integration in shrimp farming system and inclusion in government policies for amalgamation of biorights and environmental conservation.

References

Adame, M. F., Connolly, R. M., Turschwell, M. P., Lovelock, C. E., Fatoyinbo, T., Lagomasino, D., Goldberg, L. A., Holdorf, J., Friess, D. A., Sasmito, S. D., Sanderman, J., Sievers, M., Buelow, C., Kauffman, J. B., Bryan-Brown, D., & Brown, C. J. (2021). Future carbon emissions from global mangrove forest loss. *Global Change Biology, 27*(12), 2856–2866. https://doi.org/10.1111/gcb.15571

Ahmed, N., Thompson, S., & Glaser, M. (2017). Integrated mangrove-shrimp cultivation: Potential for blue carbon sequestration. *Ambio*. https://doi.org/10.1007/s13280-017-0946-2

Alexandri, E., & Jones, P. (2008). Temperature decreases in an urban canyon due to green walls and green roofs in diverse climates. *Building and Environment, 43*(4), 480–493. https://doi.org/10.1016/j.buildenv.2006.10.055

Anderson, C. M., DeFries, R. S., Litterman, R., Matson, P. A., Nepstad, D. C., Pacala, S., … Field, C. B. (2019). Natural climate solutions are not enough. *Science, 363*(6430), 933–934. https://doi.org/10.1126/science.aaw2741

Arifanti, V. B., et al. (2019). Carbon dynamics and land use carbon footprints in mangrove converted aquaculture: The case of the Mahakam Delta, Indonesia. *Forest Ecology and Management, 432*, 17–29.

Arshad, M., Alrumman, S. A., & Eid, E. M. (2018). Evaluation of carbon sequestration in the sediment of polluted and non-polluted locations of mangroves. *Fundamental and Applied Limnology, 192*, 53–64.

Barbier, E. B., Hacker, S. D., Kennedy, C., Koch, E. W., Stier, A. C., & Silliman, B. R. (2011). The value of estuarine and coastal ecosystem services. *Ecological Monographs, 81*, 169–193.

Boretti, A., & Rosa, L. (2019). Reassessing the projections of the world water development report. *NPJ Clean Water, 2*(1), 1–6. https://doi.org/10.1038/s41545-019-0039-9

Bosma, R. H., Nguyen, T. H., Siahainenia, A. J., Tran, H. T. P., & Tran, H. N. (2014). Shrimp-based livelihoods in mangrove silvo-aquaculture farming systems. *Reviews in Aquaculture, 8*(1), 43–60. https://doi.org/10.1111/raq.12072

Bowler, D. E., Buyung-Ali, L., Knight, T. M., & Pullin, A. S. (2010). Urban greening to cool towns and cities: A systematic review of the empirical evidence. *Landscape and Urban Planning, 97*(3), 147–155. https://doi.org/10.1016/j.landurbplan.2010.05.006

Bush, S. R., van Zwieten, P. A., Visser, L., van Dijk, H., Bosma, R., de Boer, W. F., & Verdegem, M. (2010). Scenarios for resilient shrimp aquaculture in tropical coastal areas. *Ecology and Society, 15*(2). https://doi.org/10.5751/es-03331-150215

Donato, D. C., Kauffman, J. B., Murdiyarso, D., Kurnianto, S., Stidham, M., & Kanninen, M. (2011). Mangroves among the most carbon-rich forests in the tropics. *Nature Geoscience, 4*(5), 293–297. https://doi.org/10.1038/ngeo1123

Food and Agriculture Organization of the United Nations. (2007). *The state of food and agriculture 2007: Paying farmers for environmental services (FAO Agriculture Series)*. FAO.

Forest Survey of India (Ministry of Environment Forest and Climate Change). (2019). STATE OF FOREST REPORT 2019. Forest Survey of India (Ministry of Environment Forest and Climate Change) Kaulagarh road, P.O. IPE Dehradun – 248195, Uttarakhand. India. https://fsi.nic.in/forest-report-2019?pgID=forest-report-2019

Goldberg, L., Lagomasino, D., Thomas, N., & Fatoyinbo, T. (2020). Global declines in human-driven mangrove loss. *Global Change Biology, 26*(10), 5844–5855. https://doi.org/10.1111/gcb.15275

Griscom, B. W., Busch, J., Cook-Patton, S. C., Ellis, P. W., Funk, J., Leavitt, S. M., … Worthington, T. (2020). National mitigation potential from natural climate solutions in the tropics. *Philosophical Transactions of the Royal Society B, 375*(1794), 20190126. https://doi.org/10.1098/rstb.2019.0126

Jayanthi, M., Thirumurthy, S., Muralidhar, M., & Ravichandran, P. (2018). Impact of shrimp aquaculture development on important ecosystems in India. *Global Environmental Change, 52*, 10–21. https://doi.org/10.1016/j.gloenvcha.2018.05.005

Kabisch, N., Frantzeskaki, N., Pauleit, S., Naumann, S., Davis, M., Artmann, M., … Bonn, A. (2016). Nature-based solutions to climate change mitigation and adaptation in urban areas: Perspectives on indicators, knowledge gaps, barriers, and opportunities for action. *Ecology and Society, 21*(2). https://doi.org/10.5751/ES-08373-210239

Kauffman, J. B., et al. (2018). Shrimp ponds lead to massive loss of soil carbon and greenhouse gas emissions in northeastern Brazilian mangroves. *Ecology and Evolution, 8*, 5530–5540.

Lewis, S. L., Wheeler, C. E., Mitchard, E. T., & Koch, A. (2019). Regenerate natural forests to store carbon. *Nature, 568*(7750), 25–28. https://doi.org/10.1038/d41586-019-01026-8

Liu, W., Chen, W., & Peng, C. (2014). Assessing the effectiveness of green infrastructures on urban flooding reduction: A community scale study. *Ecological Modelling, 291*, 6–14. https://doi.org/10.1016/j.ecolmodel.2014.07.012

Lovelock, C. E., & Duarte, C. M. (2019). Dimensions of blue carbon and emerging perspectives. *Biology Letters, 15*(3), 20180781. https://doi.org/10.1098/rsbl.2018.0781

Miao, W., & Wang, W. (2020). *Trends of aquaculture production and trade: Carp, tilapia, and shrimp*. Asian Fisheries Science, 33S. https://doi.org/10.33997/j.afs.2020.33.s1.001

Munang, R., Andrews, J., Alverson, K., & Mebratu, D. (2014). Harnessing ecosystem-based adaptation to address the social dimensions of climate change. *Environment: Science and Policy for Sustainable Development, 56*(1), 18–24. https://doi.org/10.1080/00139157.2014.861676

Primavera, J. (2006). Overcoming the impacts of aquaculture on the coastal zone. *Ocean and Coastal Management, 49*(9–10), 531–545. https://doi.org/10.1016/j.ocecoaman.2006.06.018

Primavera, J. H., Garcia, L. M. B., Castan˜os, M. T., & Surtida, M. B. (Eds.). (2000). *Mangrove-friendly aquaculture: Proceedings of the workshop on mangrove-friendly aquaculture*. SEAFDEC, Organized by the Aquaculture Department.

Primavera, J. H., Altamirano, J. P., Lebata, M. J. H. L., Delos Reyes, J. A. A., & Pitogo, C. L. (2007). Mangroves and shrimp pond culture effluents in Aklan, panay is., Central Philippines. *Bulletin of Marine Science, 80*(3), 795–804. http://docserver.ingentaconnect.com/deliver/connect/umrsmas/00074977/v80n3/s21.pdf?expires=1631537848&id=0000&titleid=10983&checksum=6395A5E31C518A7F49EEA70AA1E01E2B

Ragavan, P., Sivakumar, K., Jayaraj, R. S. C., Mohan, P. M., & Rana, T. S. (2019). Carbon storage potential of mangroves–are we missing the boat? *Current Science, 116*(6), 889–891.

Rejeki, S., Debrot, A. O., den Brink, A. M., Ariyati, R. W., & Lakshmi Widowati, L. (2020). Increased production of green mussels (Pernaviridis) using longline culture and an economic comparison with stake culture on the north coast of Java, Indonesia. *Aquaculture Research, 52*(1), 373–380. https://doi.org/10.1111/are.14900

Seddon, N., Chausson, A., Berry, P., Girardin, C. A., Smith, A., & Turner, B. (2020). Understanding the value and limits of nature-based solutions to climate change and other global challenges. *Philosophical Transactions of the Royal Society B, 375*(1794), 20190120. https://doi.org/10.1098/rstb.2019.0120

Smith, A. C., Harrison, P. A., Soba, M. P., Archaux, F., Blicharska, M., Egoh, B. N., … De Echeverria, V. W. (2017). How natural capital delivers ecosystem services: A typology derived from a systematic review. *Ecosystem Services, 26*, 111–126. https://doi.org/10.1016/j.ecoser.2017.06.006

Thirumalai, K., DiNezio, P. N., Okumura, Y., & Deser, C. (2017). Extreme temperatures in Southeast Asia caused by El Niño and worsened by global warming. *Nature Communications, 8*(1), 1–8. https://doi.org/10.1038/ncomms15531. www.nature.com/naturecommunications

Tran, N.H. (2005). Effects of mangrove leaf litters on the integrated mangrove – shrimp farming systems in Ca Mau Province, Vietnam. PhD thesis Asian Institute of Technology, Thailand.

UNFCCC. (2011). *The Cancun Agreements: outcome of the work of the Ad Hoc Working Group on Longterm Cooperative Action under the convention*. Decision 1/CP.16. COP16, FCCC/CP/2010/7/Add.1. See https://unfccc.int/resource/docs/2010/cop16/eng/07a01.pdf.

Vaiphasa, C., De Boer, W. F., Skidmore, A. K., Panitchart, S., Vaiphasa, T., Bamrongrugsa, N., & Santitamnont, P. (2007). Impact of solid shrimp pond waste materials on mangrove growth and mortality: A case study from Pak Phanang, Thailand. *Hydrobiologia, 591*(1), 47–57.

Veldman, J. W., Overbeck, G. E., Negreiros, D., Mahy, G., Le Stradic, S., Fernandes, G. W., … Bond, W. J. (2015). Where tree planting and forest expansion are bad for biodiversity and ecosystem services. *Bio Science, 65*(10), 1011–1018. https://doi.org/10.1093/biosci/biv118

Woroniecki, S. (2019). Enabling environments? Examining social co-benefits of ecosystem-based adaptation to climate change in Sri Lanka. *Sustainability, 11*(3), 772. https://doi.org/10.3390/su11030772

Chapter 8
Biorights in Pandemic Crisis

Priyanie Amerasinghe, Dipayan Dey, and Amitava Aich

8.1 Introduction

It is now presumed that the viral pandemic COVID-19 that brought the whole world to a halt might have resulted from environmental degradation, loss of ecosystem services and biodiversity or as an extended spin off effect of climate change (Damani, 2020). At least the impacts of degraded environmental aspects like air pollution and complex aerosols that could carry the pathogen had made people more susceptible to the infection. This has raised an alarm about the right of people to safe clean and healthy environment in some or the other form. The elements that substantially build up this right for the commons, are water hygiene and sanitation, clean ambient air, healthy food and all other nature's goods and services that may assure resilience to health hazards and reduce the risk of zoonosis as well. UN Human Rights Committee has endorsed that environmental degradation is the most pressing and serious threats to the ability of present and future generations to enjoy the right to life and protecting the human right to life depends on measures taken by all stakeholders to protect the environment. The global response to COVID-19 therefore, should respect, protect and fulfil rights to a healthy environment for all, leaving no one behind.

The COVID-19 pandemic has compelled us to revisit the man-animal conflict in the anthropocene. The fact that more than 65–70% infectious diseases in human are zoonotic, explains how human encroachment and urbanization has fragmented the habitat and barged into the nature. On average, one new infectious disease in humans emerges in every four months. It is important to introspect, how changing human

P. Amerasinghe
International Water Management Institute, Colombo, Sri Lanka
e-mail: p.amerasinghe@cgair.org

D. Dey · A. Aich (✉)
South Asian Forum for Environment, Kolkata, West Bengal, India

D. Dey, J. Roy (eds.), *Biorights*, Studies in Ecological Economics 7,
https://doi.org/10.1007/978-3-030-91503-2_8

behaviour and practices have disoriented the socio-ecological production landscapes and seascapes disintegrating the ecosystem goods and services that lays the foundation of human health and development. In the given scenario, a more holistic response to the COVID-19 pandemic would be integrating the human rights aspects with healthy environment in every environmental policy and agreement such as the post-2020 Global Biodiversity Framework etc. It is now crucial to reconceptualize the relationship between man and ecosystem so that the risk of health hazards from ecosystem degradation can be well prevented in the future.

The impact of environmental degradation and climate change, habitat and biodiversity loss and deterioration of ecosystem services on health and hygiene has been worst upon the marginal communities. The climate vulnerable communities, indigenous communities, migrants and environmental refugees have been the most affected groups in this pandemic, both due to the disease itself and economic downsizing as well. The impact had hit disproportionately generating new inequalities, health and economic challenges and compromising with the sustainable development goals for the communities. The pandemic had ravished shelter and food security, livelihood security and other basic amenities for the commons. In the Indian ecoregion of South Asia, the pandemic crisis was quadrupled with a number of environmental disasters during the same period, namely super cyclone Amphun (2020), Super Cyclone Taute and Yaas (2021) that furthered the explicit correlation between the nexus of poverty, disaster risk and environment. In the milieu of climate change and health vulnerability in the pandemics, therefore, the right-based approach to environment and livelihood is considered a panacea for the marginal commons.

This definitely reminds that short term economic gains derived from the deregulation of conservation objectives, always have a long-term cost. The present pandemic must, therefore, need us to reconsider the policy frameworks and practices that this grim situation owes to, rather than criticizing the decision-making system or the unpreparedness of the communities. The need of the time is to partner and step-up environmental protection and enforcement system in augmenting community resilience and preparedness through a local onus of the community that cam always be generated from a right-based approach for conservation. This can well be a future shield to protect any such disease occurrence or disaster event that may risk the life and livelihood of commons. The constitutional and legislative frameworks of States at local administrative level shall have to recognize the right to a safe, clean, healthy and sustainable environment, along with effective remedies for violations of this community right.

Integrated initiatives in protecting wildlife by preventing illegal trade can reduce potential transmission of zoonotic infections and as well promote alternative and sustainable livelihood opportunities too by enhancing tourism fees to compensate the opportunity costs. However, in the pandemic period, tourism industry was one among the worst hit businesses as since the travel and transportation was totally closed during lockdown period. On the other hand funds and grants for preventing poaching, illegal wildlife trade and other forms of prohibitions against exploitation of natural resources were all in a jeopardy bringing increased pressure on the nature-based systems. Environmental human rights defenders are essential allies in efforts to protect the environment and, by extension, human health during the COVID-19 crisis. Action is needed to protect both the environment and its defenders including, in many cases, Indigenous Peoples, whose worldviews and traditional knowledge can bring critical perspectives for sustainable and rights-based development. Limitations on civic space undermine the crucial advocacy of environmental human rights defenders, which in turn can pave the way for short-sighted and dangerous actions. Defenders should be empowered and protected from threats, reprisals, and harassment, including as relating to emergency decrees and legislation.

Community and stakeholder participation and access to information are also considered as human rights by the International Covenant on Civil and Political Rights and other international human rights instruments like Rio Principle 10, the Paris Agreement, the Aarhus Convention and the Escazù Agreement. These all agree that Governments and businesses should be transparent in sharing relevant information related to their efforts to address environmental and health crises and ensuring the informed participation of all persons in decision-making processes that affect them. New ways and modalities of working now needs to be found out. Meanwhile, Environmental governance should be modernized, including thorough inclusive and rights-based tools for digital participation and access to information, ensuring that essential environmental decision-making continues in an inclusive and effective manner regardless of the exigencies posed by COVID-19.

Free and pre informed consultation with the stakeholders, communities and beneficiaries for consideration of human rights in environmental actions as well rights-based actions for nature conservation also leads to more effective, equitable and inclusive development. Participation from all verticals and cross sections of the society, especially women and girls, local community leaders and indigenous people, offers important insights for inclusive and sustainable development, as it needs expertise of all people, their inputs and action. It is presumed that the pandemic will catalyse further democratization of decision support systems at all levels with a prime focus on conservation through inclusive consultative processes. Period of lockdown and medical emergencies have radically enhanced the use non-eco-friendly materials that include medical supplies, RTPCR testing kits and protective equipment, as well as packaging and delivery supplies made of single use plastics etc. This would eventually raise environmental hazards owing to the absence of effective and comprehensive waste management, including medical, household and other hazardous waste. It would be therefore, critical to minimize possible secondary impacts on health and the environment caused by the COVID-19 response, which can only be accomplished by the active participation of the beneficiaries, stakeholders and community-based local institutions, who can actively endure the stewardship.

The poorest, most vulnerable and marginalized communities without access to waste management or sanitation infrastructure have been, and will continue to be, hit the hardest by secondary effects on health, livelihood and rights. Preventing environmental harm and ensuring the full and effective implementation of basic human rights such as those to health, a healthy environment, and water and sanitation, is critical to prevent and minimize the risk of infectious diseases. States and other duty-bearers should ensure the safe handling and disposal of waste as a vital component of an effective and comprehensive emergency response and treat waste management, including of medical, household and other hazardous waste, as an urgent and essential public service. Effective and equitable management of biomedical and health-care waste should be guaranteed through appropriate identification, collection, separation, storage, transportation, treatment, protection, training and disposal.

A rights-based approach to the COVID-19 recovery and response requires that we build back better and more sustainably. Economic stimulus packages should protect and benefit the most vulnerable while advancing efforts to fulfil human rights, achieve the 2030 Agenda and the SDGs, and limit global heating to the greatest extent possible. The response to the crisis presents an opportunity to support improved social protection measures, and a just transition to a sustainable, no-carbon economy founded on renewable energy, environmentally sound technology, sustainable resource use, community empowerment and livelihoods of dignity. States should work jointly and individually to mobilize the maximum available resources toward building back better. Country-level socioeconomic impact analysis of COVID-19, the Common Country Analysis, UN Sustainable Development Cooperation Frameworks and the UN Secretary-General's Call to Action for Human Rights are important entry points for building back better and for operationalizing the human right to a healthy environment. The rights of all people to benefit from science and its applications must also be safeguarded ensuring that solutions to global problems, like a vaccine for COVID-19 or environmentally sound technologies, are equitably shared by all. Over the long run, inclusive, sustainable and equitable economies are more robust. All States have an obligation to pursue development that benefits both people and the planet and equitably distribute the benefits thereof. Businesses have a responsibility to respect human rights and it is also in their best interest to pursue sustainable development.

In the face of global risks, rapid, evidence-based, participatory and collective action not only produces the best results, it is also fulfilling human rights obligations. Effective responses to COVID-19 and environmental crises should be global responses grounded in solidarity, compassion, respect for human dignity and ecological integrity. The required actions and international cooperation must build on obligations of States and other duty-bearers in international legal frameworks and instruments such as the Universal Declaration on Human Rights, the Declaration on the Right to Development, and the Rio Declaration. Collaboration between governments, international partners, civil society, activists, the private sector, and all individuals and peoples are needed to fulfil human rights, including rights to a safe, clean, healthy and sustainable environment, and to achieve sustainable development that equitably meets the needs of present and future generations.

How do biorights interventions have shown resilience in this period of crisis is a very interesting study and in the following paragraphs we present a case study on the biorights intervention in some periurban Wetlands, herein.

8.2 Wetland's Biorights Intervention in Pandemics

Wetlands provide a plethora of nature's goods and services, constituting unique socio-ecological production landscapes (SEPLS) assuring food, water and livelihood security, especially for the poor. However, (Osinuga & Oyegoke, 2019) observed that these SEPLS are still excluded from mainstream conservation and are facing serious threats owing to anthropogenic pressure in the milieu of climate change. This has led to habitat destruction, biodiversity loss and deterioration of ecosystem services widening the poverty trap for the commons (Bassi et al., 2014) acknowledge that Indian peri-urban wetlands are degrading faster due to both biotic and abiotic threats like drainage and landfill, over-exploitation of fish resources, discharge of industrial effluents, uncontrolled siltation and weed infestation, ill-effects of fertilizers, pesticides and detergents. Reports in India Water Portal reveal that the South Asian ecoregion is losing 25 km^2 wetlands per square kilometer of urban area (Bansal, 2020), which is even faster than losing forest cover (Reddy et al., 2018). One-third of wetlands in the Indian subcontinent have already been severely degraded, warranting policies for management of this critical ecosystem. In the context of climate change, the mitigation potentials of wetlands are very high, though urban planning for conservation of wetlands in smart cities are meagrely addressed (McInnes, 2013).

Interventions pertaining to *'Bio-rights of Commons'* in two such peri-urban Ramsar wetlands of eastern India viz. East Kolkata Wetlands (EKW) and DeeporBeel Wetland (DBW) has proved to be a resilient and sustainable in the pandemic crisis. These wetlands were on the brink of getting removed from the Ramsar list owing to the deteriorating habitat health and biodiversity loss (Ghosh & Das, 2020; Saikia, 2019). After the successful implementation of Bio-rights project, both these two wetlands have better supported the livelihoods of communities dependent on the Ecosystem Services (ESs) of these two wetlands, ensured wetland habitat restoration and prevention of biodiversity loss, simultaneously sustainable intensification of ESs have substantially contributed to health benefits for the commons during the pandemic.

Till 1830, the tidal flow from Bay of Bengal through rivers and rivulets drenched the area with saline water, but in the latter half of the nineteenth century, as the rivers dried out due to siltation, the area started receiving the urban wastewater and local communities began wastewater fish farming in the area (Ghosh & Sen, 1987). Since then, local indigenous communities have been traditionally managing the sewer flow to regulate the nutrient load, thereby leveraging the natural resource recovery system of EKW for intensive fisheries and organic farming (Ghosh, 2005). Almost 200 years of stewardship of the wetlanders in EKW has assured them natural rights over the array of ecosystem services it provides to support livelihoods and the wetland environment relating to provisioning of wastewater recycling, food and fish production, flood protection, etc. The keepers have ensured that the habitat continues to harbour a rich and wide range of biodiversity of fishes, birds, reptiles, amphibians and plants as well (Ghosh, 1990).

The other one, DBW, an ox-bow lake of Brahmaputra River, situated in close proximity to the metro city of Guwahati in north-eastern India, is a typical wetland ecosystem within the Burma Monsoon Forest Biogeography region that provides various goods and services to the local indigenous communities, and as well, the wildlife population of the Rani-Garbhanga reserved forests, which is adjacent to the wetland and houses nearly 120 Asiatic elephants (Mitra, 2014). DBW is traditionally managed as a major fish breeding and nursery ground by three indigenous fisher's communities for supplying fish stocks to other nearby waterbodies, while conserving the local fish biodiversity. Almost 750 households in 14 fringe villages are directly dependent on the wetland for fishing and collection of herbaceous plants (Mahadevia, 2017).

Both of these wetlands are declared Ramsar sites for its wise-use practices in the anthropocene. Biophysical and socio-economic studies have been carried out and community-based management plans for the wetlands have been proposed under the purview of the biorights intervention. However, the current approach to management by the state stakeholders appears fragmented and innovative ways were found to be essentially needed to conserve and protect the ecosystem services of the wetlands (Mukherjee, 2011). Following this, a thorough institutional analysis and their responses to demands arising from impacts on wetland ecosystem services due to urban expansion was undertaken to design the biorights intervention that may support large number of marginal community members who depend on the ecosystem services of the wetland and are still excluded from the decision-making processes of the SEPLS (Dey, 2008).

This rights-based approach to poverty alleviation and conservation of the SEPLS reviews suggested that Bio-rights, as a community-based neo-fiscal conservation paradigm for protecting the ecosystems and the communities living in vicinity, being dependent on the environmental goods and services of the ecosystem can be suitable, since biorights is a sustainable neo-financing mechanism where the poor can be empowered to protect the ecosystem services that they depend upon. It was envisaged that a better institutional management framework to include bio-rights, can be appropriately developed to safeguard the ecosystem services of the wetlands, facing the wrath of urbanization.

Post-ante studies on impact assessment showed that during pandemics too, both wetland SEPLs served for the regulating services such as, maintaining local climate and air quality, carbon sequestration, moderation to extreme events such as flood and draught, erosion prevention and maintaining soil quality, pollination etc. The services were little enhanced because of a period of inaction and reduced interferences during countrywide lockdowns. These wetlands are the habitat of a huge number of species and also maintain genetic diversity, thus serve the habitat supporting services. Among all the four categories of services, provisioning services are the most prominent ones in these wetlands as they directly serve the basic needs of the local community as well as provide livelihood support to a huge number of populations. The major provisioning services include providing space for agriculture, aquaculture, livestock grazing, hunting; providing raw materials for fire woods, constructional materials, materials for craft; providing fresh water for domestic usages, irrigation, washing livestock, drinking water for livestock, commercial washing; and providing medicinal herbs. These services helped the marginal communities greatly during the lockdown period, especially for those who worked in the informal service sectors like waste workers, street vendors and domestic helps and most importantly the casual labourers, who retreated during the reverse exodus from other urban areas during the pandemic. Farming activities followed by water for irrigation, land for livestock grazing, and water for domestic usage. Biorights ensured equitable access to the communities in the hours of crisis, leaving no-one behind. Although the cultural services very much exist in the peri-urban wetlands, they meagrely attributed to the local community, during pandemics. The details of all available ecosystem services are enunciated in the following figure (Fig. 8.1).

Recent sociometric studies and FGDs showed that the Ecosystem Services (ESs) of the peri-urban wetlands were of huge importance to the peri-urban community and the communities in the downstream as well during the pandemics. Aquaculture has been the most important livelihood supporting ecosystem service of the peri-urban wetlands in EKW, availed by nearly 75% of the households. This provided cash flow in pandemics and ensured protein security in meals for the marginal, which was one of the essentials for immunity. This was followed by 48.75% of

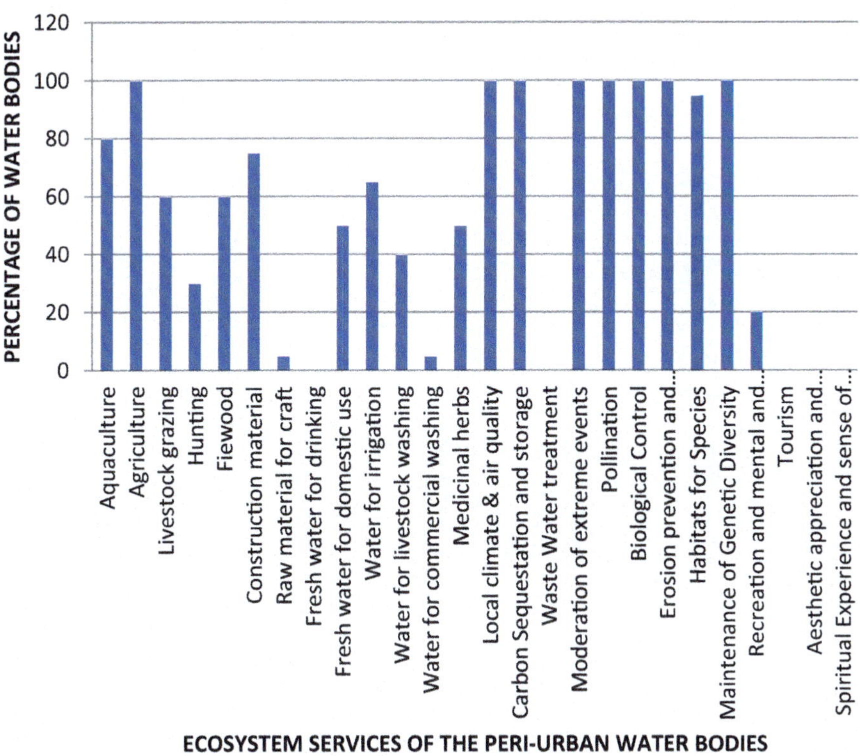

Fig. 8.1 Major ecosystem services of the peri-urban water bodies

households, who used it as agricultural space or used it as source of water for irrigation, whereas 43.75% used it as land for livestock grazing. The share of the peri-urban households collecting traditional medicines and nutritional food from the wetland areas were 81.25%. This increased almost fourfold in the peak of the second wave of infection as supporting medical services were almost unavailable. Eastern and north-eastern part of India was hit by two super cyclones, namely *Amphun* (June 2020) and *Yaas* (June 2021) and during that period 63.75% of the households were found to collect fire wood, while 43% of the households collected constructional materials, mainly wood and mud, from the wetland areas. A comparative details of these usages in pre and post pandemics are depicted in Fig. 8.2.

It has been found in this assessment that gender equity in accessing the ESs of the wetlands depends on the nature of the use of the service. Females are the predominant users of those water bodies which are mainly used for domestic purposes and small-scale fish farming at household level. On the other hand, farming activities in water bodies that are used for commercial fisheries are mostly dominated by male members. While asked to rank the importance of the water bodies, 73.75% of the respondents reported that the water bodies are of prime importance in their day-to-day life as they support their livelihood during the lockdown and fulfilled the basic needs as well. Again 68.75% of the respondents reported that the wetlands saved them from exposures to infection as they didn't have to get out for their livelihood. Thus, the studies reveal that the peri-urban wetlands SEPLS are still of prime

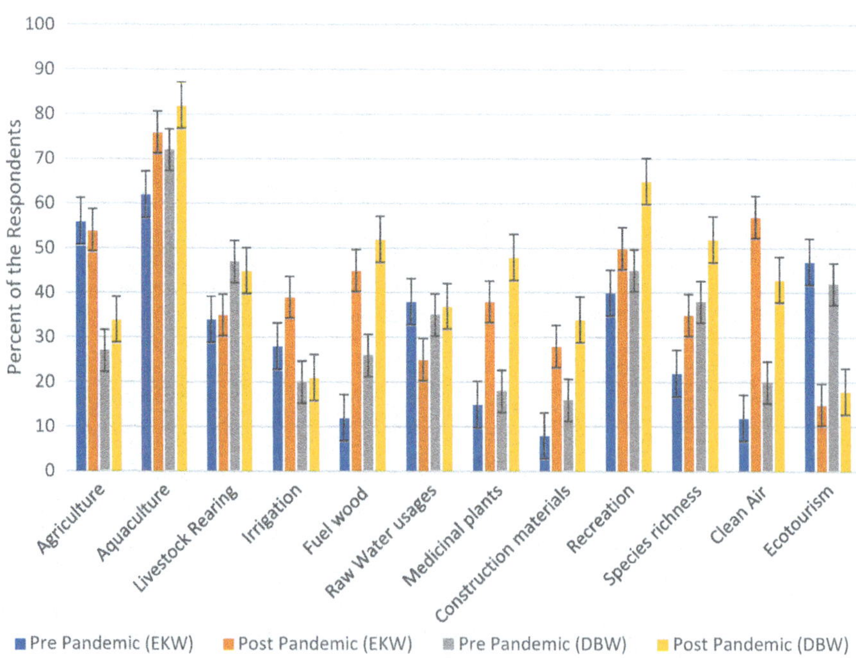

Fig. 8.2 Dependency on ecosystem services provided by the peri-urban waterbodies

importance for the peri-urban community in augmenting their resilience to disaster and crisis. These are the 'Life Line' of the peri-urban population.

8.2.1 Vulnerability and Sustainability Factors

To assess the vulnerability and sustainability of the project during pandemics, pre, post and during monsoon surveys were conducted in 2020–2021, which revealed that Livelihood Vulnerability Index (LVI) that ranges between 0 and 1, the highest vulnerability is in the urban fringes of both SEPLs wetlands. In EKW it is 0.526 and in DBW it is 0.545. In EKW, vulnerability is lowest in food security component and much low in social network component. This is because of intensive culture fisheries and paddy cultivation in EKW, whereas in DeeporBeel, only catch fisheries and livestock rearing supported livelihood. The literacy rate in households near to the urban fringe is higher than the literacy rates in far off households. Thus, beneficiaries near urban settings had better access to societal networking and were less vulnerable to pandemic panic, than those who are settled far from the urban areas, as they received regular updates through social networking and were more aware about protection and prevention of the infection. Staying near urban settlements also assure better access to water supply and health services than staying away, thereby reducing vulnerability. This prompted a trend amongst marginal households to temporarily shift near urban areas for better access to COVID relief, health services and amenities during the pandemic crisis, which led to anthropogenic encroachments and deterioration of a few wetland services, further augmenting their vulnerability. The survey also suggested that urban fringes have more pollution, water crisis and inequitable access to services and amenities for the urban poor compared to periurban and rural areas, wherein the quality of goods and services though are inferior. Detailed analysis prompted that if the dependency on nature services is high, the socioeconomic vulnerability will be high in areas near to urban settlements. Vulnerability was also found to be high if the head of the families were female or illiterate, as well for families having members without any skills. Female heads are less aggressive in securing the amenities or services owing to societal dominance of male, while illiterate and unskilled heads were incapable in acquiring the opportunities in a competitive environment owing to incompetency. More dependence on agriculture for livelihood, migration of male members due to increasing LUP changes and conversion of agricultural to non-agricultural land and more proximity to the urban areas seemingly increased the societal vulnerability, as in the case of DBW. Adopting more than one agricultural livelihood is one of the adaptive strategies for agricultural communities in these SEPLs against climate vulnerability and pandemic crisis. Thus, families having multiple agricultural practices have been considered as less vulnerable, this is why EKW showed up with a better LVI score over DBW. Community's livelihood vulnerability has been found to be related to the degree of drift of people from their traditional livelihood, as in case of EKW. Most of the families therein having at least one member in non-agricultural livelihood, or working in unorganized sectors in the urban areas, suffered more during the lockdown period. Many of the farmers practice farming in leased lands and commute to this place from other parts of the peri-urban places in both Kolkata & Guwahati. They also suffered economic losses during lockdown, which couldn't be compensated through payments of wetland services. Among the temporary settlers in wetland areas during pandemics, percentage of households having no landownership was highest in DBW, this has been a trend observed before in days of crisis (Darshini, 2017), whereas casual labourers who migrated back joined the fishers groups in EKW.

8.3 Bio-rights: Impact Assessment

The right-based neo-economic conservation model was initiated in these wetlands in the year 2016–2017 to recognize the rights of the marginal fishers and farmers in the wetland socio-ecological production landscape and capacities were built so that they get engaged in conservation and maintenance of wetland habitat, like maintenance of water quality, sewer inflow controls for regulation of nutrients leading to eutrophication, plantation of wetland species for phytoremediation and stabilization of embankments of waterbodies in the wetlands, as well maintenance of the environmental flow of water. To compensate the opportunity costs of these beneficiaries, they were bank linked and financially included through micro-insurance and microfinance so that they can leverage from the training they underwent for availing alternative livelihood opportunities. This had been a unique support during the pandemics, as they could benefit from the insurance coverages and as well could be converged easily to the government help schemes through their existing bank accounts. The following table [Table 8.1] on micro-insurance coverages availed by the beneficiaries gives a glimpse of the same. The beneficiaries were also made into Joint Liability Groups for strengthening the local institutions and promoting women entrepreneurship, which had helped them access the schemes and benefits during pandemic from governments, private sector partners and development agencies.

These supports and facilities, in recognition of their efforts and engagements, compensated their opportunity costs from payments of environment services (PES) thereby assuring livelihood security, financial inclusion and risk mitigation. Conservation of SEPLs was prompted through waterbody rejuvenation, embankment stabilization, plantation of wetland species and abatement of anthropogenic encroachments. Formation of Joint Liability Groups (JLGs) comprising of the fishers and farmers of the wetland communities, was initiated for ensuring shared responsibility and strengthened local institutional framework.

Financial inclusion of these JLGs through bank linkage, credit linkage and micro-insurance coverage spread the risk and as well ensured economic security. Gender equity and women empowerment were core components of the intervention. Review and analysis of available empirical evidences, data and information, shows striking impacts of biorights intervention in both the Ramsar Wetland SEPLs (Amerasinghe & Dey, 2018). The significant impacts are highlighted below here:

Table 8.1 Status Microinsurance Scheme: Biorights Project in East Kolkata Wetlands

Status of microinsurance cover under biorights scheme CHEME in East Kolkata Wetlands							
Sl No	Name of fishers cooperative	No of members	No of members insured	No of female members insured	No of claims till 2021 (M)	No of claims till 2021 (F)	% profit contributed by Bheri co-op
1	Sukantanagar Bheri	72	72	11	1	0	25
2	Natar Bheri	169	145	16	4	1	18
3	Gopeshwar Bheri	84	79	12	3	0	20
4	Sahebmara Bheri	227	199	33	7	0	15
5	Narkeltala Bheri	246	232	18	3	1	10
6	Sardar Bheri	180	172	19	7	0	20
7	Paresher Bheri	343	265	26	4	3	20
8	Jhograshiser Bheri	75	73	8	1	1	25
9	Chalker Bheri	82	75	12	3	0	25
10	Munshir Bheri	110	90	8	1	2	15
11	Molla Bheri	170	170	16	3	1	10
12	Daktar Bheri	160	158	15	0	0	30
13	Goltala Bheri	220	218	29	0	2	30
14	Chhoinavi Bheri	235	219	32	2	1	20
15	Patrabad Bheri No 1	352	333	45	9	6	25
16	4 No Bheri	265	262	28	1	3	20
17	Captain Bheri	227	221	36	6	3	20
18	Ghosher Bheri	56	45	6	0	0	15
19	Bantaala Bheri	165	118	11	2	2	25
20	Mahishbathan Bheri	87	80	9	0	1	15

(a) Strengthening of local institutions and enabling risk coverage through micro-insurance and financial inclusion has substantially lowered the vulnerability in three years of time span.
(b) Community-led conservation in both the SEPLS, viz. EKW and DBW, accentuated sustainable intensification of ecosystem services and that increased the returns on revenues, as evidenced by an increase of per capita income of fishers by 17–20% per year. Bio-rights intervention also leveraged alternative economic opportunities from ecotourism and circular economic interventions for women entrepreneurship.
(c) Collective responsibility in wise use practices and shared efforts for sustainable intensification of ecosystem services were more distinct in EKW, owing to the

organized patterns of community engagements and robust local institutions. This is thus considered as an indicator for sustainability, accomplished through Bio-rights.

(d) Pro conservation activities in the SEPLS improved habitat health and increased primary production in the area.

8.3.1 Assessment Studies in Pandemic Crisis

During COVID pandemic, nationwide lockdown was announced, triggering mass reverse exodus of temporary migrants working as casual labourers and there was intense loss of livelihood in the informal sector. During such crisis, nearly 2000 landless laborers belonging to the wetland communities returned back to EKW and DBW. The Bio-rights intervention could minimize the adversity both on life and livelihood of these migrants without compromising with the conservation objectives. The intervention supported these migrants with some economic opportunities through manual labour work in farming and fisheries, and engagement in supply chains of the farm sectors. While the intervention extended health support to the urbanites living in the near vicinity by providing protein supplements in daily food from table fish, fox nuts and duckery products, production of fresh vegetables and local medicinal plants supported indigenous communities in wetlands during the crisis. The agro-biodiversity preserved in the local cropping cycles, which are mostly managed by women farmers in the wetland SEPLS are sources of indigenous food and food supplements for augmenting health and immunity of the local people. This had a great bearing on the indigenous life during the pandemic crisis. The urbanites in the vicinity, who visited the wetland eco-tourism trails for morning walk and physical warm-up, during the 'work-from-home' period in pandemics, experienced the ambience of fresh air, oxygenated and free from urban pollution as a source of health condiment. An attitude scaling on the impacts of health outcomes of these eco-trails is indicated in the info-graph below (Fig. 8.4) that shows 50–60% agreed about rebuilding immunity and getting fresh air from the SEPLS, while 37% admitted the protein food supplements made available from these wetlands as important (Fig. 8.3).

8.3.2 Outcomes & Impact

Wetlands, though alienated from mainstream conservation owing to deficient policy frameworks, are unique socio-ecological production landscapes and seascapes which are significant in the milieu of climate change. This paper reviews the SEPLS of two peri-urban Ramsar sites to put forward ecosystem based adaptive strategies for conservation and recommend a framework using right-based neo-economic approach to compensate the opportunity costs of the wetland inhabitants and

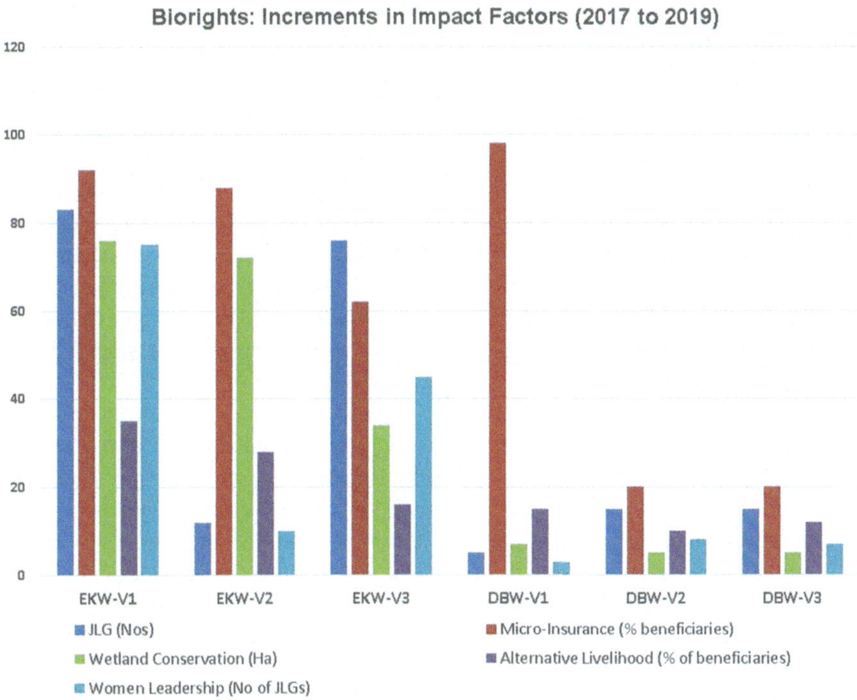

Fig. 8.3 Biorights: increments in impact factors

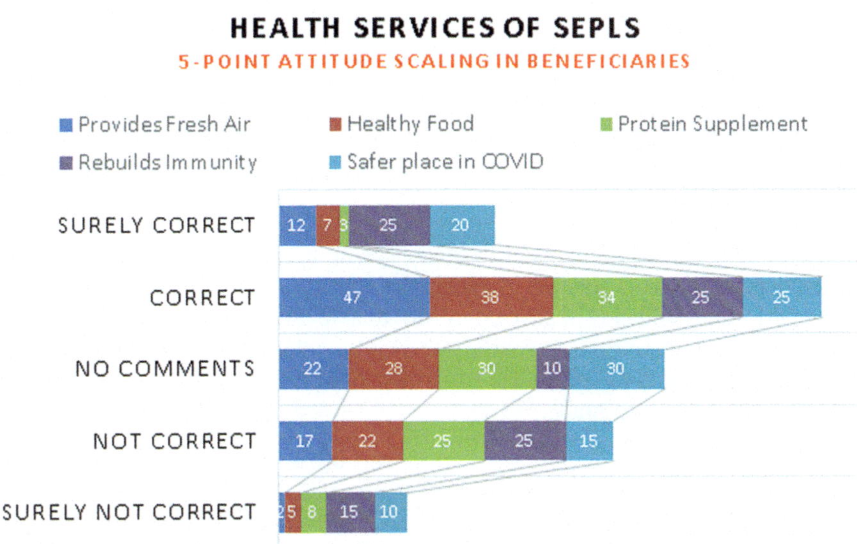

Fig. 8.4 Attitude scaling on health services of SEPLS

dependents. This paper makes a case for this approach based on the following observations.

(a) It is observed that vulnerable communities in these SEPLS continually explore the opportunities amidst challenges imposed on them, as has been during the pandemic and they attempt to reduce the impact of externalities through appropriate resource stocks, extraction and marketing strategies. Since most of the beneficiaries work in informal sectors, wherein institutional arrangements are somehow fragile and cannot tackle onslaughts imposed by large scale and systemic perturbations, the beneficiaries either chose to accept working as casual labourers in urban areas or change to alternative economic opportunities in urban service deliveries as evidenced from the sociometric assessments in the wetland communities.
(b) Perusal of outcomes from focused group discussions and household survey in the SEPLS reveals that under economic pressure, responsibilities of risk mitigation move from individual to collective action enabling the sharing of economic losses by suitably changing livelihood options. However, mechanisms to monitor compliance and enforcement of the norms for collective community action are not easily practised due to the diverse perception of risks. It's obvious that the compensation of loss and damage is mostly bestowed on use of nature's goods and services. This leads to loss of habitat and biodiversity owing to overuse.
(c) Predictive analyses based on participatory evaluation of ecosystem services on the fate of wetland functions under worst-case scenarios can help largely to define the best scenarios too, in Bio-rights centred livelihood promotion coupled with community conservation strategies and the related institutional mechanisms developed in course of implementation of the Bio-rights project.
(d) By tracing the biodiversity, community dynamics and relevance of spatio-temporal changes in LULC (land use and land cover), as well institutional arrangements that have formalized with the management of the EKW and DBW over years of community-based conservations in these two SEPLS, the paper aspires to explores avenues to design a more inclusive sustainable ecosystem service-based management plan that include novel concepts like bio-rights.

Reviewing the determinants of dependences of marginal agrarian communities on wetland ecosystem services vis-à-vis the impacts of urban encroachment on these peri-urban wetlands in advocating sustainable intensification of primary productivity, a right-based conservation framework could be recommended for wetlands near urban growth centers, for adaptive policy planning. The brief overview of this framework is as hereunder.

(i) In recognizing the importance of these peri-urban wetland habitat and its services, a normative practice has to be enunciated, in measuring the economic value of its goods and services, ensuring that the estimations are non-ambiguous and non-discriminatory at all levels of policy and practice to leverage the benefits of conserving these wetlands, especially in the climate milieu.

(ii) The peri-urban wetlands must be recognized by the town planners and urban architects as socio-ecological production landscapes and the returns on investments in intensifying primary productivity in these wetlands must add to the GDP. This would enable sharing the benefits, as well co-benefits earned from goods and services of the wetlands, with the marginal communities to ensure inclusive growth.
(iii) Ecosystem-based adaptation for conservation of peri-urban wetlands needs to be included as mandatory clauses in town planning and landscaping, leaving no one behind in participatory management and be recommended for enriching national nature capital alike forests and rivers.
(iv) The lessons learned must be incorporated in the draft for developing National Wetland (conservation and management) policy in India for inclusion and adoption. The following facets may be considered for further reference in including in the draft and as well knowledge economy.

A. Reference frame for defining community-based wetland conservation strategies and enabling platform for redressal of issues pertaining to it

(i) Conservation strategies for peri-urban wetlands are to be outlined at community-ecosystem interface to recognize community governance in conservation, alike policies and practice in place as in Community Forest Management.
(ii) Local wetland authority entrusted for conservation and management of wetlands SEPLS have to be inclusive and have equitable community representations for redressal of issues as well assure remedial support to ensure equity, reciprocity and partnership for conservation.
(iii) A frame of reference is expected to be in place, to assure conservation and wise use of peri-urban wetlands through a national policy document, to be referred as the National Wetland Policy [NWP] in mitigating any or all issues pertaining to conservation and management of these wetlands.

B. Ownership and commons property rights

(i) With regard to the ownership and property rights on wetland ecosystems, it is recommended that be it private, joint and/or state owned, alterations in conservation regulations, changes pertaining to land use pattern of the SEPLS or changes in usages, it would mandatorily need to have the commons bio-rights and conservation of biodiversity as priorities to upkeep the ecosystem services for all stakeholders, ensuring equitable access and benefit sharing.
(ii) In cognizance to the above mandates, ecosystem services of these SEPLS must be recognized as commons property rights for economic co-benefits, irrespective of the ownership of the wetlands, whereas efforts for conservation would be recognized as an 'equal and reciprocal responsibility' for all facilitating partnership and participation beyond the property rights.

(iii) The conservation and management regulations and as well the commons property rights on wetlands shall not alter with the altercation and/or transfer in ownership rights.

C. Participatory conservation paradigm and Bio-rights of commons

(i) Planning procedures for conservation must encourage and facilitate as well, a participatory practice, leaving no one behind to promote inclusive partnership.
(ii) Implementation of the conservation plan must have a robust monitoring and periodic evaluation (M&E) system and feedback systems for adaptive learnings and constructive review of the plan being implemented.
(iii) In enforcing the implementation plan and considering wise use as conservation priority, the opportunity costs of commons incurred in forgoing the co-benefits of ecosystem services needs to be estimated and compensated through novel and uniform financial architypes. The compensation may either be used for risk spreading or coverage, as may be decided collectively by the participants from time to time.

D. National Wetland Policy for conservation and management

(i) The state must implicate on emergency basis the propounding of a national reference agenda for conservation and management of wetlands, to be recognized as the "National Wetland Policy" that may consider the details, as discussed above, to be integral to its operational framework.
(ii) The state must appoint efficient, trained and empathetic experts as custodians of the policy upon implementation to facilitate the conservation objectives and augment its ecosystem services thereto.
(iii) The policy framework, thus built, must seek concurrence with all national policies for conserving other potential ecosystems to undermine the conflicts of interests and opinions in overlapping jurisprudence and co-parallel regulations so as to avoid abundance and loss of national natural resources.

References

Amerasinghe, P. H., & Dey, D. (2018). *Recommendations for the wise use of urban and peri-urban wetlands in Kolkata, India*. International Water Management Institute (IWMI). CGIAR Research Program on Water, Land and Ecosystems (WLE). 8p. (WLE BRIEFING SERIES NO. 23). https://cgspace.cgiar.org/handle/10568/96987

Bansal & Swati. (2020). India water portal. Retrieved on 7 Mar 2020, from https://www.indiawaterportal.org/articles/mumbai-lost-71-percent-wetlands-last-four-decades-report

Bassi, N., Kumar, M. D., Sharma, A., & Pardha-Saradhi, P. (2014). Status of wetlands in India: A review of extent, ecosystem benefits, threats and management strategies. *Journal of Hydrology: Regional Studies, 2*, 1–19. https://doi.org/10.1016/j.ejrh.2014.07.001

Damani, A. (2020). COVID-19 and climate change. *Sushruta, 13*(2) pre-print v1; ePUB 03.05.2020. https://doi.org/10.38192/13.2.10

Darshini Mahadevia, A. M. (2017). Ecology vs housing and the land rights movement in Guwahati. *Economic and Political Weekly*.

Dey, D. (2008). Digital library of commons, Indian University. Retrieved September. 12, 2008, from https://dlc.dlib.indiana.edu/dlc/bitstream/handle/10535/1874/Dey_124901.pdf?sequence=1

Ghosh, A. K. (1990). Biological resources of wetlands of East Calcutta. *Indian Journal of Landscape System and Ecological Studies, 13*(1), 10–23.

Ghosh, D. (2005). *Ecology and traditional wetland practice: Lessons from wastewater utilisation in the East Calcutta wetlands*. Worldview.

Ghosh, S., & Das, A. (2020). Wetland conversion risk assessment of East Kolkata wetland: A Ramsar site using random forest and support vector machine model. *Journal of Cleaner Production, 275*, 123475. https://doi.org/10.1016/j.jclepro.2020.123475

Ghosh, D., & Sen, S. (1987). Ecological history of Calcutta's wetland conversion. *Environmental Conservation, 14*(3), 219–226.

Mahadevia, D., Mishra, A., & Joseph, Y. (2017). Ecology vs housing and the land rights movement in Guwahati.

McInnes, R. J. (2013). Recognising wetland ecosystem services within urban case studies. *Marine and Freshwater Research, 65*(7), 575–588. https://doi.org/10.1071/MF13006

Mitra, S., & Bezbaruah, A. N. (2014). Railroad impacts on wetland habitat: GIS and modeling approach. *Journal of Transport and Land Use, 7*(1), 15–28. https://doi.org/10.5198/jtlu.v7i1.181

Mukherjee, D. P. (2011). Stress of urban pollution on largest natural wetland ecosystem in East Kolkata-causes, consequences and improvement. *Archives of Applied Science Research, 3*(6), 443–461.

Osinuga, O. A., & Oyegoke, C. O. (2019). Degradation assessment of wetlands under different uses: Implications on soil quality and productivity. *African Journal of Agricultural Research, 14*(1), 10–17. https://doi.org/10.5897/AJAR2018.12981

Reddy, C. S., Saranya, K. R. L., Pasha, S. V., Satish, K. V., Jha, C. S., Diwakar, P. G., … Murthy, Y. K. (2018). Assessment and monitoring of deforestation and forest fragmentation in South Asia since the 1930s. *Global and Planetary Change, 161*, 132–148. https://doi.org/10.1016/j.gloplacha.2017.10.007

Saikia, J. L. (2019). Deepor Beel wetland: Threats to ecosystem services, their importance to dependent communities and possible management measure. *Natural Resources and Conservation, 7*(2), 9–24. https://doi.org/10.13189/nrc.2019.070201

Index

D. Dey, J. Roy (eds.), *Biorights*, Studies in Ecological Economics 7,
https://doi.org/10.1007/978-3-030-91503-2

D

E

F

G

The manufacturer's authorised representative in the EU is Springer Nature Customer Service Centre GmbH, Europaplatz 3, 69115 Heidelberg, Germany. If you have any concerns regarding our products, please contact ProductSafety@springernature.com

Printed and bound by CPI Group (UK) Ltd, Croydon, CR0 4YY
28/09/2026
02224077-0019